T0192511

One-Dimensional Dynamical Systems

One-Dimensional Dynamical Systems
An Example-Led Approach

Ana Rodrigues

University of Exeter, UK

CRC Press
Taylor & Francis Group
Boca Raton London New York

CRC Press is an imprint of the
Taylor & Francis Group, an **informa** business
A CHAPMAN & HALL BOOK

First edition published 2022
by CRC Press
6000 Broken Sound Parkway NW, Suite 300, Boca Raton, FL 33487-2742

and by CRC Press
2 Park Square, Milton Park, Abingdon, Oxon, OX14 4RN

© 2022 Ana Rodrigues

CRC Press is an imprint of Taylor & Francis Group, LLC

Library of Congress Cataloging-in-Publication Data

Names: Rodrigues, Ana (Mathematics), author.
Title: One-dimensional dynamical systems : an example-led approach / Ana Rodrigues, University of Exeter, UK.
Description: First edition. | Boca Raton : Chapman & Hall/CRC Press, 2021. | Includes bibliographical references and index.
Identifiers: LCCN 2021019440 (print) | LCCN 2021019441 (ebook) | ISBN 9780367701109 (hardback) | ISBN 9780367701086 (paperback) | ISBN 9781003144618 (ebook)
Subjects: LCSH: Differentiable dynamical systems.
Classification: LCC QA614.8 .R655 2021 (print) | LCC QA614.8 (ebook) | DDC 515/.39--dc23
LC record available at https://lccn.loc.gov/2021019440
LC ebook record available at https://lccn.loc.gov/2021019441

ISBN: 9780367701109 (hbk)
ISBN: 9780367701086 (pbk)
ISBN: 9781003144618 (ebk)

DOI: 10.1201/9781003144618

Contents

CHAPTER 1

Introduction

For almost every phenomenon from physics, chemistry, biology, medicine, economics and other sciences one can make a mathematical model that can be regarded as a dynamical system.

Our goal is to understand the eventual or asymptotic behaviour of an iterative process.

If this process is a differential equation whose independent variable is time, then the theory aims to predict the behaviour of solutions of the equation when either $t \to \infty$ or $t \to -\infty$.

When studying discrete dynamical systems, we are interested in investigating the sequence of iterates

$$\underbrace{x_0, f(x_0), f(f(x_0)), \dots.}_{n \text{ times}}$$

The main questions we are going to deal with are whether the system exhibits chaotic behaviour or whether it is periodic.

Let us start by introducing some basic concepts that we will use throughout this book.

We say a point x_f is a *fixed point* for our function f if $f(x_f) = x_f$.

The *forward orbit* of a point x is the set of points $x, f(x), f^2(x)$, $f^3(x), \dots$ (note that here $f^n(x)$ denotes the nth-iterate of our function) and we usually denote this orbit by $\mathcal{O}^+(x)$. When f is a homeomorphism we can define the *full orbit* of our point x as the set of points $f^n(x)$ for $x \in \mathbb{Z}$. The same way we can think about the *backward orbit* of our point x and we denote it by $\mathcal{O}^-(x)$.

An important concept in dynamical systems are the periodic points. A point x is said periodic of period n if $f^n(x) = x$. The least positive

n for which this relation holds is called the *prime period* of x. We will refer to the set of periodic points of period n by $\text{Per}_n(f)$.

We can also talk about *eventually periodic* points, this occurs when there exists a natural number $m > 0$ such that $f^{n+i}(x) = f^i(x)$ for all $i \leq m$) (in a sense $f^i(x)$ is periodic of period n for $i > m$).

Finally, we say that a point x is forward asymptotic to p if

$$\lim_{t \to \infty} f^{in}(x) = p.$$

Here it is important to note that the stable set of p is the set of all points that are forward asymptotic to p and we usually designate it by $W^s(p)$.

Let us see what happens to the map $f(x) = x^2 - 1$ (see Figure 1.1). To find the fixed points, we must solve the equation $f(x) = x$, that is, the quadratic equation $x^2 - 1 = x$ which gives us the fixed points $x = 1 \pm \sqrt{5}/2$. Now, we have $f(-1) = 0$ and $f(f(-1)) = f(0) = -1$ so it is clear that -1 is a periodic point for f of period 2.

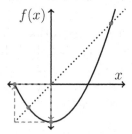

Figure 1.1: Plot of the map $f(x) = x^2 - 1$. The diagonal has been drawn in to illustrate the positions of the fixed points at $x = (1 \pm \sqrt{5})/1$ (that lie along the diagonal) and the path of the periodic points as a web diagram $x = -1, 0$ under iteration.

Another important concept is the one of the *critical point*. These are the points where the first derivative vanishes, this is, the points x such that $f'(x) = 0$. If we consider again the function $f(x) = x^2 - 1$, we see that $f'(x) = 2x$ and so, 0 is a critical point for this function. A critical point x is said to be non degenerate if $f''(x) \neq 0$ and degenerate when $f''(x) = 0$.

Our main aim is to study the orbits of points under given maps. This is not an easy task. A good example is the study *on iterations of* $1 - ax^2$ *in* $(-1, 1)$ by Benedicks and Carleson in 1985.

Examples are, in a sense, one of the most important things when studying Mathematics. In this book, we will study a particular family of maps and some of the work recently done in order to study its dynamics.

In 1965, V. Arnold investigated the family of *standard maps* of the circle given by

$$A_{a,b}(x) = x + a + \frac{b}{2\pi}\sin(2\pi x) \pmod 1,$$

here by "mod 1" we mean that both the arguments and the values are taken modulo 1. This family of maps is also known as *Arnold maps* and it was important in creating KAM theory [3].

Standard maps have applications in biology, in particular in the study of the dynamical system of a beating heart [4] where the standard map is used to explain the effects of brief periodic electrical stimulation on embryonic chick heart cells. Arnold tongues, which we will define in detail later, have been used to study why heart-beats and respiration synchronise in a resting or anaesthetised patient [13].

In [27–29] the family of *double standard maps* given by

$$f_{a,b}(x) = 2x + a + \frac{b}{\pi}\sin(2\pi x) \pmod 1 \tag{1.1}$$

was investigated.

For $b = 1$, maps of the family (1.1) have a unique cubic critical point (at $c = 1/2$) and negative Schwarzian derivative. Thus, they behave similarly to the quadratic maps. For $b < 1$, there is no critical point, so the maps are local diffeomorphisms.

Complexification of the maps, obtained by conjugacy via $e^{2\pi i x}$, gives the family

$$g_{a,b}(z) = e^{2\pi i a} z^2 e^{b\left(z - \frac{1}{z}\right)}. \tag{1.2}$$

Those maps are symmetric with respect to the unit circle and factored by this symmetry, they have only one critical point and no asymptotic values in $\mathbb{C} \setminus \{0\}$. Therefore a map $f_{a,b}$ has at most one attracting or neutral periodic orbit (see [18, 27]). This conjugacy will be studied in detail.

One can also look at the family of double standard maps as a hybrid between the family of *standard maps* and *expanding maps* of the circle.

Some recent work has been done for classes of families that include double standard maps. Misiurewicz and Rodrigues studied them in [27, 28]. Benedicks and Rodrigues [5] investigated symbolic dynamics for this family. Universality for critical circle covers was studied by Levin and Świątek, [20]. Levin and van Strien [21] proved complex bounds, quasisymmetric rigidity and density of hyperbolicity for a class of real analytic maps which includes the double standard maps. Fagella and Garijo [18] studied a class of complex maps containing the maps $g_{a,b}$.

As for the Arnold's family, for the double standard family, we call the sets for which there is an attracting periodic orbit of a given type (period plus combinatorics) *tongues*. Dezotti [16] proved that tongues are connected. The lowest tongue tip is at $b = 1/2$, for the period 1 tongue. If $0 < b < 1/2$, the map $f_{a,b}$ is expanding. At higher b-levels, there may be finitely or infinitely many tongues (see [27]). In particular, at the critical level $b = 1$ all tongues are present, and it is easy to prove that they are dense at this level (see [21]).

Investigation of double standard maps naturally generates an interest in the behaviour of higher degree circle maps. What happens if we add the sinusoid not to the identity or the doubling map but to the tripling map, quadrupling map or α-tupling map, i.e. the map given by $x \mapsto \alpha x$ for any $\alpha \in \mathbb{N}$?

Motivated by these questions, we get the following family of α-*standard maps*,

$$f_{\alpha,a,b}(x) = \alpha x + a + \frac{\alpha b}{2\pi} \sin(2\pi x) \ (\text{mod } 1), \tag{1.3}$$

where α is an integer.

Like with the double standard family of maps given by (1.1), for α-standard maps, we will rescale b in order to keep the critical value at 1.

This book contains background to the theory of dynamical systems and a collection of material from several sources and articles written in collaboration with different people.

This book follows work in collaboration with Robert Ashton, Michael Benedicks, Noah Cockram, Alex Pace and Michal Misiurewicz. I would like to thank for their collaboration during my career. I would like also to thanks Pete Ashwin for a careful look at the manuscript.

CHAPTER 2

Rotation numbers

Let us denote the unit circle by S^1. We can represent this in several ways. We can think of S^1 as being the complex numbers with absolute value 1 or as being $\mathbb{R}/\mathbb{Z}$ so that $S^1 = [0,1]$, the set of real numbers taken modulo 1. We will most often consider S^1 in the latter form. Consider the map $\pi : \mathbb{R} \to S^1$ which is the restriction of the real numbers to the circle. For this reason, it can take different forms depending on how we choose to denote the circle. We can use exponential notation, where $\pi(x) = \exp(2\pi i x)$ which represents a rotation of x radians. Alternatively, we can think of circle maps as the real numbers taken mod 1. In this way, $\pi(x) = x \pmod 1$ and we can think of, say $t \in S^1$, as representing a rotation of $2\pi t$ radians. We will most often use the second approach. We say π is a *covering map* as it wraps $\mathbb{R}$ around S^1 [15].

Definition 2.0.1. A continuous smooth map $f : S^1 \to S^1$ of the circle is *expanding* if $|f'(x)| > 1$ for all $x \in S^1$. Furthermore, if $f'(x) > 1$ then we say f is *orientation preserving*. If $f'(x) \leq 1$ then we say f is *orientation reversing*.

Definition 2.0.2. We define the *lift* $F : \mathbb{R} \to \mathbb{R}$ of an orientation preserving homeomorphism $f : S^1 \to S^1$ to be a map satisfying $\pi \circ F = f \circ \pi$ such that the following conditions hold:

$$F \text{ is monotonically increasing} \tag{2.1}$$

$$F(t+1) = F(t) + 1 \text{ for all } t \in \mathbb{R}. \tag{2.2}$$

Note that for any homeomorphism f, if we have two lifts F_1, F_2 then there exists $k \in \mathbb{Z}$ such that $F_2(t) = F_1(t) + k$ for all t and in fact k is independent of t (see for instance [15]). The rotation number is defined as follows.

Definition 2.0.3. For an orientation preserving homeomorphism f we define the *rotation number of f* in terms of the lift F of f. Let,

$$\rho_0(F, t) = \lim_{n \to \infty} \frac{F^n(t) - t}{n},$$

we define the *rotation number of f* to be,

$$\rho(f) = \rho_0(F, t) \pmod{1}.$$

Let us now prove that this limit exists. Before this let us introduce some tools.

Definition 2.0.4. We define the *ceiling function* of a number x to be the function $\lceil x \rceil : \mathbb{R} \to \mathbb{Z}$ such that $\lceil x \rceil$ is equal to the smallest integer z such that $x \leq z$. We can see from the definition that $x \leq \lceil x \rceil < x + 1$, we will use this fact in the next lemma.

Lemma 2.0.5. *For any two points $t, s \in \mathbb{R}$ there exists $l \in \mathbb{Z}$ such that $t \leq s + l < t + 1$.*

Proof. We use the idea of the ceiling function introduced above to prove this lemma. Define $l = \lceil t - s \rceil$ such that $t - s \leq l < t - s + 1$. Add s to the inequality to get $t - s + s \leq l + s < t - s + 1 + s$, i.e. $t \leq l + s < t + 1$ as required. $\qquad\square$

We introduce the idea of lim inf and lim sup here as we will need it in order to prove that rotation numbers exist.

Definition 2.0.6. Let $\{S_n\}_{n \geq 1}$ be a bounded real sequence, so there exists $M_1, M_2 \in \mathbb{R}$ with $M_1 \leq s_n \leq M_2$ for all $k \geq 1$. Consider the sequence

$$\alpha_k = \sup\{s_n : n \geq k\}.$$

We can see from the definition of α_k that the sequence $\{\alpha_k\}$ is decreasing, that is, $\alpha_{k+1} \leq \alpha_k$ for all k and it is bounded below by M_1. Thus $\{\alpha_k\}_{k \geq 1}$ converges towards the infimum of its range. We define,

$$\limsup_{n \to \infty} s_n = \lim_{k \to \infty} \alpha_k = \inf_{k \geq 1} \sup_{n \geq k} s_n.$$

Similarly, we define a sequence for $k \geq 1$,

$$\beta_k = \inf\{s_n : n \geq k\}.$$

We can see that $\{\beta_k\}$ is an increasing sequence, $\beta_{k+1} > \beta_k$, and that it is bounded above by M_2. Thus $\{\beta_k\}_{k\geq 1}$ converges towards the supremum of its range. We define,

$$\liminf_{n\to\infty} s_n = \lim_{k\to\infty} \beta_k = \sup_{k\geq 1} \inf_{n\geq k} s_n.$$

Proposition 2.0.7. *Let $\{s_n\}_{n\geq 1}$ be a bounded real sequence. Then,*

$$\liminf_{n\to\infty} s_n \leq \limsup_{n\to\infty} s_n.$$

Moreover, the sequence is convergent with limit L if and only if $\liminf_{n\to\infty} s_n = \limsup_{n\to\infty} s_n = L$.

Remark 2.0.8. We omit the proof of this proposition. It is useful to note that from the above proposition if we can show that $\liminf_{n\to\infty} s_n \geq \limsup_{n\to\infty} s_n$ then this implies $\liminf_{n\to\infty} s_n = \limsup_{n\to\infty} s_n$ and so is a sufficient condition to prove that s_n converges to a limit, i.e. that $\lim_{n\to\infty} s_n$ exists. We will use this proposition to prove the next theorem.

Although the proof of the next theorem is well known, we present here a detailed discussion.

Theorem 2.0.9. *Let $f : S^1 \to S^1$ be an orientation preserving homeomorphism with lift F. Then,*

a. *for $t \in \mathbb{R}$, the limit in the definition of $\rho_0(F,t)$ exists and is independent of t,*

b. *$\rho(f) = \rho_0(F,t)$ (mod 1), and this is independent of the lift chosen, and*

c. *$\rho(f)$ depends continuously on f.*

Proof. Take any two points $t, s \in \mathbb{R}$. By Lemma 2.0.5, there exists $l \in \mathbb{Z}$ such that $t \leq s + l < t + 1$. From the definition of the lift F, it is monotonic increasing and $F(t+1) = F(t) + 1$ for any $t \in \mathbb{R}$. Using the above inequality and the properties of the lift F we can write

$$F(t) \leq F(s+l) < F(t+1) = F(t) + 1.$$

We want to show that if we apply F, p times, for any integer $p \geq 1$ that this inequality still holds, namely that

$$F^p(t) \leq F^p(s+l) < F^p(t) + 1. \tag{2.3}$$

We aim to prove this result using induction. We have shown that (2.3) holds for $p = 1$.

Now we assume it is true for $p = m$ and we prove it holds for $p = m + 1$. Take $F^m(t) \leq F^m(s+l) < F^m(t) + 1$ which is our induction step and apply F to see that

$$F(F^m(t)) \leq F(F^m(s+l)) < F(F^m(t) + 1),$$

this gives

$$F^{m+1}(t) \leq F^{m+1}(s+l) < F(F^m(t)) + 1,$$

and we get

$$F^{m+1}(t) \leq F^{m+1}(s+l) < F^{m+1}(t) + 1.$$

Hence, by induction, (2.3) holds for all $p \geq 1$, $p \in \mathbb{Z}$.

Now consider the two sides of the inequality separately,

$$F^p(t) \leq F^p(s+l) \tag{2.4}$$
$$F^p(s+l) < F^p(t) + 1. \tag{2.5}$$

We subtract $t + 1$ from (2.4) to get,

$$F^p(t) - t - 1 \leq F^p(s+l) - t - 1.$$

We have that $t \leq s + l < t + 1$ so we can deduce, $-t - 1 < -s - l \leq -t$, which transforms the inequality to,

$$F^p(t) - t - 1 < F^p(s+l) - s - l. \tag{2.6}$$

Subtracting $s + l$ from (2.5) we can see that

$$F^p(s+l) - s - l < F^p(t) + 1 - s - l,$$

and again using the fact that $-t - 1 < -s - l \leq -t$ we can write,

$$F^p(s+l) - s - l < F^p(t) + 1 - t \tag{2.7}$$

Note that

$$F^p(s+l) = F^p((s+l-1) + 1) = F^p(s+l-1) + 1.$$

Hence, by induction, as l is an integer, $F^p(s+l) - s - l = F^p(s) - s$. We now substitute this into inequality (2.6) and (2.7) to get,

$$F^p(t) - t - 1 < F^p(s) - s, \tag{2.8}$$
$$F^p(s) - s < F^p(t) - t + 1. \tag{2.9}$$

Let $s = F^n(t)$ we can write

$$F^{p+n}(t) - t = F^p(F^n(t)) - F^n(t) + F^n(t) - t = F^p(s) - s + s - t.$$

Adding $s - t$ to inequality (2.9) we get

$$F^p(s) - s + s - t < F^p(t) - t + 1 + s - t$$

and substituting into this our expression for $F^{p+n}(t) - t$ we get the inequality,

$$F^{p+n}(t) - t < F^p(t) - t + 1 + F^n(t) - t. \tag{2.10}$$

Let, $p = n$ and substitute into inequality (2.10) to get

$$F^{2p}(t) - t < F^p(t) - t + 1 + F^p(t) - t.$$

Collecting like terms gives, $F^{2p}(t) - t < 2[F^p(t) - t] + 1$. We assume the statement holds for power kp, i.e.

$$F^{kp}(t) - t < k[F^p(t) - t] + k - 1.$$

We aim to prove it holds for power $(k + 1)p$ then by induction it holds for all integers k greater than or equal to 2.

We write

$$F^{(k+1)p}(t) = F^{kp+p}(t).$$

Substituting this into (2.10) we get

$$F^{kp+p}(t) - t < F^{kp}(t) - t + 1 + F^p(t) - t.$$

Using our induction step gives

$$F^{kp+p}(t) - t < k[F^p(t) - t] + k - 1 + 1 + F^p(t) - t$$

and collecting like terms we see $F^{kp+p}(t) - t < (k + 1)[F^p(t) - t] + k$ which is precisely what we wanted to show. Hence, by induction for any $k \in \mathbb{Z}, k \geq 2$ we have,

$$F^{kp}(t) - t < k[F^p(t) - t] + k - 1. \tag{2.11}$$

Euclidean division states that for any positive integers μ, η such that $\mu \geq \eta$ we can find $a, b \in \mathbb{Z}$, with $a \geq 0$ and $0 \leq b < \eta$ such that $\mu = a\eta + b$. So here for any $n \geq p$, as n, p are positive integers we can use Euclidean division to write $n = kp + i$ where $0 \leq i < p$ and we write $F^n(t) - t = F^{kp+i}(t) - t$.

Applying (2.10) we see that

$$F^n(t) - t < F^{kp}(t) - t + F^i(t) - t + 1.$$

Now we use (2.11) to see that

$$F^n(t) - t < k[F^p(t) - t] + F^i(t) - t + k.$$

Dividing this by n and using the fact the $n \geq kp \Rightarrow \frac{1}{n} \leq \frac{1}{kp}$ we get,

$$\frac{F^n(t) - t}{n} < \frac{k[F^p(t) - t] + F^i(t) - t + k}{n}$$

$$= \frac{k[F^p(t) - t] + k}{n} + \frac{F^i(t) - t}{n}$$

$$\leq \frac{k[F^p(t) - t] + k}{kp} + \frac{F^i(t) - t}{n}$$

$$= \frac{k[F^p(t) - t]}{kp} + \frac{F^i(t) - t}{n} + \frac{k}{kp}$$

This simplifies to give an upper bound for $\frac{F^n(t)-t}{n}$,

$$\frac{F^n(t) - t}{n} < \frac{F^p(t) - t}{p} + \frac{F^i(t) - t}{n} + \frac{1}{p}. \tag{2.12}$$

In a similar way we find a lower bound for $\frac{F^n(t)-t}{n}$. From (2.8) we have $F^p(t) - t - 1 < F^p(s) - s$. Again let $s = F^n(t)$ and substitute into the inequality, we see,

$$F^{p+n}(t) - t > F^p(t) - t - 1 + s - t. \tag{2.13}$$

Letting $n = p$ and collecting like terms gives that $F^{2p}(t) - t > 2[F^p(t) - t] - 1$ and like before we can use induction to show that,

$$F^{kp}(t) - t > k[F^p(t) - t] - k + 1. \tag{2.14}$$

Let $n = kp + i$, then combining (2.13) and (2.14) we can write,

$$F^{kp+i}(t) - t > F^{kp}(t) - t - 1 + F^i(t) - t$$

$$> k[F^p(t) - t] - k + 1 - 1 + F^i(t) - t$$

$$= k[F^p(t) - t] + F^i(t) - t - k.$$

Like before we divide both sides of

$$F^{kp+i}(t) - t > k[F^p(t) - t] + F^i(t) - t - k$$

by n and use the fact that $(k+1)p > n$ implies $\frac{1}{(k+1)p} < \frac{1}{n}$ to write,

$$\frac{F^n(t) - t}{n} > \frac{k[F^p(t) - t] + F^i(t) - t - k}{n}$$

$$= \frac{k[F^p(t) - t] - k}{n} + \frac{F^i(t) - t}{n}$$

$$> \frac{k[F^p(t) - t] - k}{(k+1)p} + \frac{F^i(t) - t}{n}.$$

Which gives the lower bound,

$$\frac{k[F^p(t) - t]}{(k+1)p} + \frac{F^i(t) - t}{n} - \frac{k}{(k+1)p} < \frac{F^n(t) - t}{n}. \tag{2.15}$$

Taking the inequalities (2.12) and (2.15) together we have found upper and lower bounds for $\frac{F^n(t)-t}{n}$. If we fix p we can consider the sequence $\sup\{\frac{F^j(t)-t}{j} : j \geq n\}$. Clearly our bounds for $\frac{F^n(t)-t}{n}$ hold for this sequence too, we have

$$\frac{k[F^p(t) - t]}{(k+1)p} + \frac{F^i(t) - t}{n} - \frac{k}{(k+1)p} \leq \sup\left\{\frac{F^j(t) - t}{j} : j \geq n\right\}$$

$$\sup\left\{\frac{F^j(t) - t}{j} : j \geq n\right\} \leq \frac{F^p(t) - t}{p} + \frac{F^i(t) - t}{n} + \frac{1}{p}.$$

As $n \to \infty$, $k \to \infty$ because $n = kp + i$ with p fixed and $0 \leq i \leq p$. Hence taking the limit as $n \to \infty$ gives bounds for $\limsup \frac{F^n(t)-t}{n}$; the upper bound (2.12) becomes,

$$\lim_{k \to \infty} \left(\frac{F^p(t) - t}{p} + \frac{F^i(t) - t}{n} + \frac{1}{p}\right) = \frac{F^p(t) - t}{p} + \frac{1}{p}$$

and the lower bound (2.15) becomes,

$$\lim_{k \to \infty} \left(\frac{k[F^p(t) - t]}{(k+1)p} + \frac{F^i(t) - t}{n} - \frac{k}{(k+1)p}\right).$$

By L'Hôpital's rule, we have that this limit is simply given by

$$\lim_{k \to \infty} \left(\frac{F^p(t) - t}{p} + \frac{F^i(t) - t}{n} - \frac{1}{p} \right)$$
$$= \frac{F^p(t) - t}{p} - \frac{1}{p}.$$

Combining our upper and lower bounds for $\limsup$ we get

$$\frac{F^p(t) - t}{p} - \frac{1}{p} \le \limsup_{n \to \infty} \frac{F^n(t) - t}{n} \le \frac{F^p(t) - t}{p} + \frac{1}{p}, \qquad (2.16)$$

indicating that $\limsup \frac{F^n(t) - t}{n}$ is finite. We now let p vary and consider the sequence as p goes to infinity of $\inf \{ \frac{F^j(t) - t}{j} : j \ge p \}$. We can see that our lower bound from above still applies, namely that

$$\limsup \frac{F^n(t) - t}{n} \le \inf \left\{ \frac{F^j(t) - t}{j} : j \ge p \right\}$$

and taking the limit as $p \to \infty$ the bound still applies, giving us the fact that,

$$\limsup \frac{F^n(t) - t}{n} \le \liminf \frac{F^p(t) - t}{p}.$$

By Proposition (2.0.7), this tells us that

$$\lim_{n \to \infty} \frac{F^n(t) - t}{n}$$

exists and is finite for any $t \in \mathbb{R}$. This limit is the definition of the rotation number $\rho_0(F, t)$, so we have shown that the rotation number exists, it remains to prove that it is independent of t.

Dividing inequalities (2.8) and (2.9) by p gives,

$$\frac{F^p(t) - t - 1}{p} < \frac{F^p(s) - s}{p} < \frac{F^p(t) - t + 1}{p}.$$

We now take the limit as p goes to infinity to see that,

$$\lim_{p \to \infty} \frac{F^p(t) - t - 1}{p} \le \lim_{p \to \infty} \frac{F^p(s) - s}{p} \le \lim_{p \to \infty} \frac{F^p(t) - t + 1}{p}$$
$$= \lim_{p \to \infty} \frac{F^p(t) - t}{p} \le \lim_{p \to \infty} \frac{F^p(s) - s}{p} \le \lim_{p \to \infty} \frac{F^p(t) - t}{p}$$
$$= \rho_0(F, t) \le \rho_0(F, s) \le \rho_0(F, t),$$

which means that $\rho_0(F, t) = \rho_0(F, s)$ for any two points $t, s \in \mathbb{R}$, i.e. rotation number is independent of the point. We can denote rotation number as just $\rho_0(F)$.

We now prove (b). Above we stated that for two different lifts F_1, F_2 of f there exists $k \in \mathbb{Z}$ such that $F_2(t) = F_1(t) + k$. To prove that $\rho_0(f)$ is independent of the lift F we use the fact that, $F_2^n(t) = F_1^n(t) + nk$ so we need to show this holds by using induction on $n \geq 1$. We have shown it holds for the base case and we assume it holds for $n = m$. We try to show it holds for $n = m + 1$,

$$
\begin{aligned}
F_2^{m+1}(t) &= F_2^m(F_2(t)) \\
&= F_1^m(F_2(t)) + mk \\
&= F_1^m(F_1(t) + k) + mk \\
&= F_1^m(F_1(t)) + k + mk \\
&= F_1^{m+1}(t) + (m+1)k.
\end{aligned}
$$

Hence by induction, $F_2^n(t) = F_1^n(t) + nk$. We substitute our expression for F_2^n into the definition for rotation number (2.0.3) to see that,

$$
\begin{aligned}
\rho_0(F_2) &= \lim_{n \to \infty} \left(\frac{F_1^n(t) - t}{n} + \frac{nk}{n} \right) \\
&= \rho_0(F_1) + k.
\end{aligned}
$$

We have $k \in \mathbb{Z}$ so therefore, $\rho_0(F_2) = \rho_0(F_1) \pmod 1$.

Finally, we prove (c). We aim to show that the rotation number is continuously dependent on f. Let $\epsilon > 0$, to show continuous dependence we want to show there exists a $\delta > 0$ such that if g is another homeomorphism of the circle which is close enough to f in terms of the C^0 topology, i.e. if $|f - g| < \delta$ then $|\rho(f) - \rho(g)| < \epsilon$.

We have shown earlier that the rotation number does not depend on the point, so we will choose here to define the rotation number in terms of the point $t = 0$,

$$
\rho_0(F) = \lim_{n \to \infty} \frac{F^n(0)}{n}. \tag{2.17}
$$

Let F be the lift of f. We can find an integer p such that $p \leq F^n(0) < p + 1$ hence we can write $p - 1 < F^n(0) < p + 1$. We can choose $\delta > 0$ such that g has a lift G satisfying, $p - 1 < G^n(0) < p + 1$.

Apply F^{nk} times to 0 and notice that we can rewrite it as,

$$
\begin{aligned}
F^{nk}(0) &= F^{nk}(0) - 0 \\
&= F^n(F^{(k-1)n}(0)) - F^{(k-1)n}(0) + \cdots \\
&\quad + F^n(F^{jn}(0)) - F^{jn}(0) + \cdots + F^n(F^n(0)) - F^n(0) \\
&\quad + F^n(F^0(0)) - F^0(0) \\
&= \sum_{j=1}^{k-1} \left[F^n(F^{jn}(0)) - F^{jn}(0) \right].
\end{aligned}
$$

We see that writing $F^{nk}(0)$ in this way consists of k summations and for each $j = \{0, 1, \ldots, k-1\}$, by writing $t = F^{jn}(0)$ and using the fact that $\exists p \in \mathbb{Z}$ such that $p - 1 < F^n(t) - t < p + 1$ we see that,

$$
p - 1 < F^n(F^{jn}(0)) - F^{jn}(0) < p + 1.
$$

Combining these k inequalities gives that

$$
k(p-1) < \sum_{j=1}^{k-1} \left[F^n(F^{jn}(0)) - F^{jn}(0) \right] = F^{nk}(0) < k(p+1). \quad (2.18)
$$

We can write $G^{nk}(0)$ in the exact same way,

$$
G^{nk}(0) = \sum_{j=1}^{k-1} \left[G^n(G^{jn}(0)) - G^{jn}(0) \right],
$$

to give that,

$$
k(p-1) < G^{nk}(0) < k(p+1). \quad (2.19)
$$

Dividing inequalities (2.18) and (2.19) by nk and taking the limit as $k \to \infty$ gives,

$$
\frac{p-1}{n} < \lim_{k \to \infty} \frac{F^{nk}(0)}{nk} < \frac{p+1}{n}
$$

and

$$
\frac{p-1}{n} < \lim_{k \to \infty} \frac{G^{nk}(0)}{nk} < \frac{p+1}{n}.
$$

From our definition of rotation number (2.17) we have,

$$
\frac{p-1}{n} < \rho_0(F) < \frac{p+1}{n}
$$

and
$$\frac{p-1}{n} < \rho_0(G) < \frac{p+1}{n}.$$

Negating the inequality for $\rho_0(G)$ and adding to the inequality for $\rho_0(F)$ we get,

$$\frac{p-1}{n} - \frac{p+1}{n} < \rho_0(F) - \rho_0(G) < \frac{p+1}{n} - \frac{p-1}{n}.$$

Simplifying we see,

$$-\frac{2}{n} < \rho_0(F) - \rho_0(G) < \frac{2}{n}.$$

And we can conclude that,

$$|\rho_0(F) - \rho_0(G)| < \frac{2}{n}. \tag{2.20}$$

We can choose n such that $\frac{2}{n} < \epsilon$. Hence, $\rho(f)$ depends continuously on f. □

Example 2.0.10. Let

$$\tau_\omega(\theta) = \theta + 2\pi\omega,$$

which represents a translation of θ through an angle $2\pi\omega$. For each $k \in \mathbb{Z}$,

$$T_{\omega,k}(x) = x + \omega + k$$

is the lift of $\tau_\omega(\theta)$. What is the rotation number, $\rho(\tau)$ of τ?

To find the rotation number, first we need to verify that $T_{\omega,k}$ is a lift. To do this we show it satisfies the properties in Definition 2.0.2.

$$T_{\omega,k}(x+1) = x + 1 + \omega + k = T_{\omega,k}(x) + 1.$$

Hence it satisfies the first condition, we now consider $\frac{dT_{\omega,k}(x)}{dx} = 1 \geq 0$ hence $T_{\omega,k}$ is monotonic increasing. It remains to show that $\pi \circ T_{\omega,k} = \tau_\omega \circ \pi$.

$$(\pi \circ T_{\omega,k})(x) = x + \omega + k \pmod 1 = x + \omega \pmod 1$$
$$(\tau_\omega \circ \pi)(x) = x \pmod 1 + 2\pi\omega.$$

We see that $\pi \circ T_{\omega,k}$ represents a rotation of $2\pi\omega$ radians and we see that $\tau_\omega \circ \pi$ represents a rotation of $2\pi\omega$ radians so if we write in the form of the real numbers mod 1 we see that $(\tau_\omega \circ \pi)(\theta) = \theta + \omega \pmod 1$, just what we have for $\pi \circ T_{\omega,k}$ hence they are equal and hence $T_{\omega,k}$ is a lift of τ_ω.

We can now use Definition 2.17 to find the rotation number of the lift and then restrict this to the circle to find the lift of τ_ω.

We need to find an expression for $T_{\omega,k}^n(x)$, we can see that when we iterate $T_{\omega,k}$ each time we add $\omega + k$ to x. So, if we apply $T_{\omega,k}$ n times, we get $T_{\omega,k}^n(x) = x + n(\omega + k)$. Substituting this into our definition we see that,

$$\rho_0(T_{\omega,k}) = \lim_{n\to\infty} \frac{T_{\omega,k}^n(0)}{n} = \lim_{n\to\infty} \frac{n(\omega + k)}{n} = \omega + k$$

[15]. Restricting this to the real numbers mod 1 gives us the rotation number of τ_ω. Namely

$$\rho(\tau_\omega) = \omega + k \pmod 1 = \omega \pmod 1.$$

Example 2.0.11. Next, we want to find the rotation number of

$$f(x) = x + \frac{1}{4\pi} \sin(2\pi x) \pmod 1$$

with lift

$$F(x) = x + \frac{1}{4\pi} \sin(2\pi x).$$

We begin by proving this is indeed a lift. Consider $F(x+1)$ given by

$$F(x+1) = x + 1 + \frac{1}{4\pi} \sin(2\pi(x+1))$$

$$= x + \frac{1}{4\pi} \sin(2\pi x + 2\pi) + 1$$

$$= x + \frac{1}{4\pi} \left(\sin(2\pi x)\cos(2\pi) + \sin(2\pi)\cos(2\pi x) \right) + 1$$

$$= x + \frac{1}{4\pi} \sin(2\pi x) + 1$$

$$= F(x) + 1.$$

If we also consider

$$\frac{\mathrm{d}F(x)}{\mathrm{d}x} = 1 + \frac{1}{4\pi}\cos(2\pi x)2\pi$$
$$= 1 + \frac{1}{2}\cos(2\pi x).$$

So, we see that $\frac{\mathrm{d}F}{\mathrm{d}x} > 0$ hence F is monotonic increasing. It remains to show that $\pi \circ F = f \circ \pi$. We have,

$$(\pi \circ F)(x) = x + \frac{1}{4\pi}\sin(2\pi x) \pmod 1$$
$$(f \circ \pi)(x) = z + \frac{1}{4\pi}\sin(2\pi z) \text{ where } z = x \pmod 1.$$

We note that our expression for $(f \circ \pi)(x)$ can be rewritten. We know that $\sin(2\pi z) = \sin(2\pi x)$, where $z = x \pmod 1$ so we can write $(f \circ \pi)(x) = x + \frac{1}{4\pi}\sin(2\pi x) \pmod 1$ and we can see that $\pi \circ F = f \circ \pi$ hence F is indeed the lift of f.

We now want to find the rotation number using the Definition 2.17. We want an expression for $F^n(0)$. Now $F(0) = 0 + \frac{1}{4\pi}\sin(0) = 0$ so we have that $F^n(0) = F(0) = 0$. So we can find the rotation number of the lift,

$$\rho_0(F) = \lim_{n \to \infty}\frac{F^n(0)}{n} = \lim_{n \to \infty}\frac{0}{n} = 0.$$

Hence the rotation number of our circle map is, $\rho(f) = 0$.

We now state some definitions which will be useful later.

Definition 2.0.12. The *critical point* of a differentiable morphism $f : X \mapsto Y$ is a point $x \in A$ such that $\frac{\mathrm{d}f(x)}{\mathrm{d}x} = 0$.

Definition 2.0.13. A *periodic point of period n* of $f : X \mapsto X$ is a point $x \in X$ such that $f^n(x) = x$ for some $n \in \mathbb{N}$. If $n = 1$ we say that x is a *fixed point* of f. Furthermore, if f is differentiable then for any periodic point x of period n we can further qualify x. If,

1. $|(f^n)'(x)| < 1$ we say x is an *attracting* periodic point,

2. $|(f^n)'(x)| = 1$ we say x is a *neutral* periodic point,

3. $|(f^n)'(x)| > 1$ we say x is a *repelling* periodic point [15].

Definition 2.0.14. For a homeomorphism $f : X \mapsto X$ the *orbit of a point* x is the set of points,

$$O(x) = \{f^n(x)|n \in \mathbb{Z}\}$$

[15]. We can see that if x is a periodic point of f of period n then the orbit of x is finite and in fact has n elements,

$$O(x) = \left\{x, f(x), f^2(x), \ldots, f^{n-1}(x)\right\}.$$

If this is the case we call $O(x)$ the *periodic orbit of x of period n*.

Lemma 2.0.15. *The rotation number of an orientation preserving homeomorphism f is an integer if and only if it has a fixed point.*

Proof. Assume f has a fixed point. We want to show that the rotation number of f is an integer. Let $x_0 \in S^1$ be our fixed point, so $f(x_0) = x_0$ and let $p \in \mathbb{R}$ be such that $\pi(p) = x_0$, where $\pi(x)$ is the natural projection of $\mathbb{R}$ onto S^1.

Now let F be a lift of f. Recall that $\pi \circ F = f \circ \pi$ so as $(f \circ \pi)(p) = f(\pi(p)) = x_0$ we have that $(\pi \circ F)(p) = x_0$ which gives that $F(p) = \pi^{-1}(x_0) = p + m$ for some $m \in \mathbb{Z}$.

As F is a lift it is true that $F(x+1) = F(x) + 1$.

We assume $F(x+m) = F(x) + m$. This gives

$$F(x+m+1) = F(x+m) + 1 = F(x) + m + 1$$

hence, by induction on m we have $F(x+m) = F(x) + m$ for $m \geq 1$. So we can see that

$$F^2(p) = F(p+m) = F(p) + m = p + 2m.$$

We assume $F^n(p) = p + nm$, so

$$F^{n+1}(p) = F(F^n(p)) = F(p+nm) = p + (n+1)m.$$

Hence by induction on n, we get $F^n(p) = p + nm$, for $n \geq 1$. Now using the definition of rotation number we have that,

$$\rho_0(F) = \lim_{n \to \infty} \frac{F^n(p)}{n} = \lim_{n \to \infty} \frac{p + nm}{n} = m.$$

Which is an integer and $\rho(f) = \rho_0(F) \pmod 1$ and so $\rho(f)$ is also an integer.

Conversely, we assume the rotation number of f is an integer and prove that f has a fixed point. Let $\widetilde{F}$ be a lift of f, we assume $\rho_0(\widetilde{F}) = m \in \mathbb{Z}$. Let $F = \widetilde{F} - m$, this is also a lift of f. We have by part (b) of the proof of (2.0.9) that $\rho_0(F) = \rho_0(\widetilde{F}) - m = 0$. We now consider the orbit $F^n(0)$, as F is a lift it is increasing so $F^n(0)$ is monotone. This is because if $F(0) \geq 0$ then assuming $F^n(0) \geq F^{n-1}(0)$ and applying F and using the fact that F is increasing we get that $F^{n+1}(0) > F^n(0)$ and so by induction on n, $F^n(0) \geq F^{n-1}(0)$ for all $n \geq 1$. A similar result follows for the case $F(0) < 0$.

Suppose for a contradiction that $F^n(0)$ is unbounded. This means in particular that $\exists n_0 \in \mathbb{N}$ such that $|F^{n_0}(0)| > 1$. Now consider

$$F^{mn_0}(0) = (F^{n_0}(0))^m = F^{n_0}(0) + F^{n_0}(0) + \cdots + F^{n_0}(0),$$

the sum of $F^{n_0}(0)$ m times. From this we can clearly see that $|F^{mn_0}(0)| > m$ for all $m \in \mathbb{N}$. We divide this inequality by mn_0 to get,

$$\frac{|F^{mn_0}(0)|}{mn_0} > \frac{m}{mn_0} = \frac{1}{n_0}.$$

We notice that this says that $|\rho_0(F)| = |\lim_{m\to\infty} \frac{F^{mn_0}(0)}{mn_0}| > \frac{1}{n_0}$, where n_0 is some finite natural number. This contradicts the fact that $\rho_0(F) = 0$ which we computed earlier. Hence $F^n(0)$ must be bounded.

We have established that $F^n(0)$ is both bounded and monotonic and therefore has a limit. Let $x_* = \lim_{n\to\infty} F^n(0)$. Applying F again gives,

$$F(x_*) = F\left(\lim_{n\to\infty} F^n(0)\right) = \lim_{n\to\infty} \left(F^{n+1}(0)\right) = x_*.$$

So,

$$F(x_*) = x_*,$$

i.e. x_* is a fixed point of F.

Let $p_* = \pi(x_*) = x_* \pmod 1$. Then as $\pi \circ F = f \circ \pi$ we have that $p_* = (\pi \circ F)(x_*) = (f \circ \pi)(x_*) = f(p_*)$. So $f(p_*) = p_*$, f has a fixed point. □

Corollary 2.0.16. *An orientation-preserving circle homeomorphism has a rational rotation number if and only if it has a periodic point. Equivalently $\rho(f) = \frac{m}{n}$ if and only if $\exists p \in S^1$ such that $f^n(p) = p$.*

Proof. We notice from the definition of rotation number (2.17) that

$$\rho(f^m) = \lim_{n\to\infty} \frac{F^{nm}(0)}{n},$$

$$= \lim_{n\to\infty} \frac{(F^n(0))^m}{n},$$

$$= \lim_{n\to\infty} \left(\underbrace{\frac{F^n(0)}{n} + \frac{F^n(0)}{n} + \cdots + \frac{F^n(0)}{n}}_{m \text{ times}} \right),$$

$$= m \lim_{n\to\infty} \frac{F^n(0)}{n},$$

$$= m\rho(f).$$

The corollary follows from this result. We have that

$$\rho(f) = \frac{m}{n}$$

if and only if

$$n\rho(f) = m,$$

this is,

$$\rho(f^n) = m.$$

Thus, by (2.0.15), there exists $p \in S^1$ such that $f^n(p) = p$.

$\square$

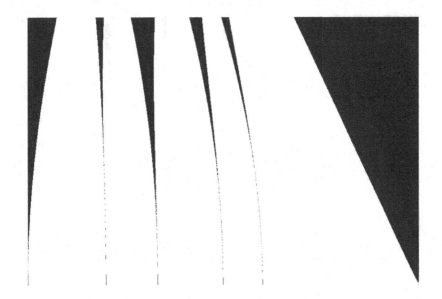

Figure 2.1: Arnold tongues for the family of standard maps.

2.1 ARNOLD TONGUES FOR DOUBLE STANDARD MAPS

Let us look at some computer generated pictures (see [27]) regarding the family of double standard maps

$$f_{a,b}(x) = 2x + a + \frac{b}{\pi}\sin(2\pi x) \pmod 1,$$

and also the Arnold family of maps

$$A_{a,b}(x) = x + a + \frac{b}{2\pi}\sin(2\pi x) \pmod 1.$$

It is common, when studying the standard (Arnold) family of maps, to show the parameter values for which there is an attracting periodic orbit in the (a,b)-plane. These are the so-called *Arnold tongues* (see Figure 2.1). The parameter values for which there is an attracting periodic orbit are sometimes designated *phase-locking regions*.

We can easily see that (mod 1) we have $A_{-a,b}(-x) = -A_{a,b}(x)$. This means that $A_{-a,b}$ is conjugate to $A_{a,b}$ via the map $x \mapsto -x \pmod 1$. Thus, when looking at the picture of the Arnold tongues we see that this picture is symmetric (with respect to the line $a = 1/2$). Usually we

Figure 2.2: Arnold tongues for the family of double standard maps.

then only show this picture for $1/2 \leq a \leq 1$. This also applies to our maps $f_{a,b}$ (replacing $A_{a,b}$).

In Figure 2.1, the vertical axis shows us b, from 0 to 1. The horizontal axis shows us a, from $1/2$ to 1.

In the picture, we show all tongues of period 5 or less, and their order beginning from left to right is 2, 5, 3, 4, 5, 1. These correspond to the rotation numbers

$$1/2 < 3/5 < 2/3 < 3/4 < 4/5 < 1/1.$$

Now let us look at pictures 2.1 and 2.2 and let us compare them. In Figure 2.2 the vertical axis is b, from $1/2$ to 1 and the horizontal axis is a, from $1/2$ to 1.

In this picture we show all tongues of period 5 or less, and their order from left to right is

$$1, 5, 5, 4, 5, 5, 4, 3, 5, 5, 4, 5, 5, 4, 3, 5, 5, 2, 5, 5, 4, 5, 3, 5, 4, 5.$$

Later we will explain that this corresponds to the rational numbers $0/1 < 1/31 < 2/31 < 1/15 < 3/31 < 4/31 < 2/15 < 1/7 < 5/31 < 6/31 < 3/15 < 7/31 < 8/31 < 4/15 < 2/7 < 9/31 < 10/31 < 1/3 <$

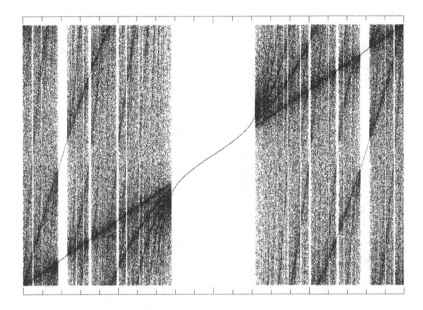

Figure 2.3: The (a, x)-plot, with a and x from 0 to 1 and $b = 1$.

$11/31 < 12/31 < 6/15 < 13/31 < 3/7 < 14/31 < 7/15 < 15/31$. Note that the denominators are of the form $2^n - 1$ (where n is the period).

This order above is completely different from the one we get for the standard maps. One of the main differences between the two classes of maps is that for double standard maps the tongues do not begin all at the level $b = 0$ (it is easy to see from the pictures that they begin at much higher levels.)

An interesting fact is that for double standard maps the lowest tongue tip is at $b = 1/2$ (for the period 1 tongue) as for $0 < b < 1/2$ the map is expanding.

It seems that for a given value of $b \in [0, 1)$ there are only finitely many tongues. However, we will later show that this is not true.

We will now show the technical machinery developed in [27] in order to explain the size of the tongues both in the a and b directions.

In order to study the a-size, we should measure it at the level $b = 1$.

Having the value of b fixed, it makes sense to see what happens to the picture in the (a, x)-plane (like the classical pictures for the family of the logistic or real quadratic families of maps). See Figure 2.3 that presents the global picture, with both a and x varying from 0 to 1.

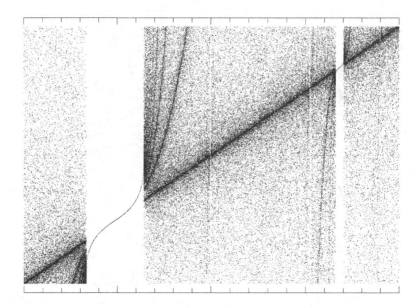

Figure 2.4: The (a, x)-plot, with a and x from 0.69053 to 0.69055 and $b = 1$.

Since $1/2$ is the unique critical point of $f_{a,1}$ and $f_{a,b}$ has negative Schwarzian derivative, we know already that this implies that for every a there is at most one attracting periodic orbit. (see, e.g. [14]). If such an orbit exists, one of its points must be close to $1/2$. Since $f_{a,1}(1/2) = a$, there must be a point of such orbit close to a.

Figure 2.3 suggests that in order to see well how the attracting periodic orbit varies with a, it is better to look close to the diagonal $x = a$, rather than close to the line $x = 1/2$, where the line is very steep unless the period is very small.

Blow-ups at many regions close to the diagonal show a graph of a periodic point as a function of a that is not so steep in its middle part (although of course it has to be vertical at the boundary of the window). This is illustrated in Figure 2.4, where a and x vary from 0.69053 to 0.69055.

However, there are periodic orbits close to the boundaries of tongues of small period, that we can call *resonant* or *intermittent*, for which this graph is much steeper. Figure 2.5 shows what happens near the boundary of the period 1 tongue. There a and x vary from 0.61087 to 0.61093.

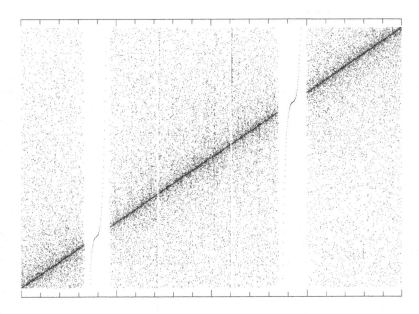

Figure 2.5: The (a, x)-plot, with a and x from 0.61087 to 0.61093 and $b = 1$.

2.2 ARNOLD TONGUES FOR α-STANDARD MAPS

Recall our family of *α-standard maps* in (1.3) given by

$$f_{\alpha,a,b}(x) = \alpha x + a + \frac{ab}{2\pi} \sin(2\pi x) \ (\text{mod } 1),$$

with $b = 1$ and $0 \leq a < 1$.

Denote the lift of $f_{a,b}(x)$ as in (2) to the real line by $F_{a,b}(x)$. For the family (1.1), in [27] semiconjugacy is established as the limit defined by

$$\Phi_{a,b}(x) = \lim_{n \to \infty} \frac{F_{a,b}^n(x)}{2^n}.$$

It is proved that this limit exists uniformly in x and that it semiconjugates $F_{a,b}$ with multiplication by 2.

We will now consider the family of circle maps given by (1.3) where the parameters $a, b \in \mathbb{R}$ with $0 \leq b \leq 1$ and $\alpha \in \mathbb{N}$ and we denote its lift to the real line by $F_{\alpha,a,b}(x)$.

We start by including graphs of $F_{\alpha,a,b}$ for $a = -0.3$, $b = 0.7$ and $\alpha = 3, 4$, see Figures 2.6 and 2.7.

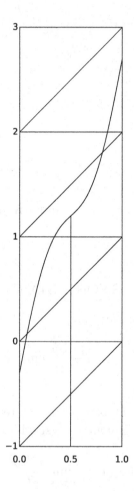

Figure 2.6: The graph of $F_{\alpha,a,b}$, when $\alpha = 3$, $a = -0.3$ and $b = 0.7$.

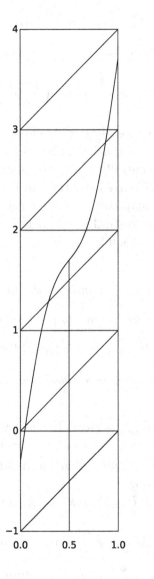

Figure 2.7: The graph of $F_{\alpha,a,b}$, when $\alpha = 4$, $a = -0.3$ and $b = 0.7$.

In [27], it was shown that there are 2 points of the double standard maps such that $F'_{a,b} = 1$, however, this is not the case for α-standard maps. We can see this by computing

$$F'_{\alpha,a,b}(x) = \alpha + \alpha b \cos(2\pi x)$$

which is equal to 1 when

$$x = \frac{1}{2\pi} \arccos\left(\frac{1-\alpha}{\alpha b}\right).$$

This only has solutions for $\left|\frac{1-\alpha}{\alpha b}\right| \leq 1$ which is when $\left|\frac{1}{\alpha} - 1\right| \leq b$. Note that in this case there is no such p for $b = 0.7$ and $\alpha \geq 4$. We can see that the graph of $F_{\alpha,a,b}$ is centrally symmetric about the point $(\frac{1}{2}, F_{\alpha,a,b}(\frac{1}{2}))$.

Figures 2.6 and 2.7 make clear the effect of taking the rotation with the α multiplication map, which is that the map maintains a similar shape but is stretched vertically; each increase in x results in a much larger increase in $F_{\alpha,a,b}$ than for double standard maps for different values of α.

Proposition 2.2.1. *$F_{\alpha,a,b}(x)$ satisfies the following properties:*

1. *$F_{\alpha,a,b}$ is continuous increasing (as a function of x),*

2. *$F_{\alpha,a,b}(x + k) = F_{\alpha,a,b}(x) + \alpha k$ for all $k \in \mathbb{Z}$,*

3. *$F_{\alpha,a,b}(x)$ is increasing as a function of a and continuous jointly in x, a, b.*

Proof. Continuity of $F_{\alpha,a,b}$ is clear. In order to show it is increasing we consider $\mathrm{d}F_{\alpha,a,b}(x)/\mathrm{d}x = \alpha + \alpha b \cos(2\pi x)$, which is clearly greater than or equal to 0 as $|\alpha b \cos(2\pi x)| \leq |\alpha|$ and as $\alpha > 0$ we have that $\frac{\mathrm{d}F_{\alpha,a,b}}{\mathrm{d}x} \geq 0$. Hence $F_{\alpha,a,b}$ is increasing.

We now prove that $F_{\alpha,a,b}(x + k) = F_{\alpha,a,b}(x) + \alpha k$. We have

$$F_{\alpha,a,b}(x + k) = \alpha(x + k) + a + \frac{\alpha b}{2\pi} \sin(2\pi(x + k))$$

$$= \alpha x + a + \frac{\alpha b}{2\pi} (\sin(2\pi x) \cos(2\pi k) + \sin(2\pi k) \cos(2\pi x)) + \alpha k$$

$$= F_{\alpha,a,b}(x) + \alpha k.$$

We now prove property 3, continuity is clear. In order to see that $F_{\alpha,a,b}(x)$ is increasing as a function of a, note that $\mathrm{d}F_{\alpha,a,b}(x)/\mathrm{d}a = 1$. $\square$

For the family of double standard maps, the limit $\Phi_{a,b}$ as above can be used to classify the tongues. In the next lemma we generalize this limit for the α-standard family:

Lemma 2.2.2. *Let $F_{a,b}(x) = F_{\alpha,a,b}(x)$ be the lift to the real line of (1.3). The limit*

$$\Phi_{\alpha,a,b}(x) = \lim_{n\to\infty} \frac{F_{a,b}^n(x)}{\alpha^n} \qquad (2.21)$$

exists uniformly in x. Also $\Phi_{\alpha,a,b}(x)$ is an increasing function of x which satisfies $\Phi_{\alpha,a,b}(x + k) = \Phi_{\alpha,a,b}(x) + k$ for every $k \in \mathbb{Z}$ and $\Phi_{\alpha,a,b}(F_{a,b}(x)) = \alpha\Phi_{\alpha,a,b}(x)$ for every x. So $\Phi_{\alpha,a,b}$ semiconjugates $F_{a,b}$ with multiplication by α.

Thus, $\Phi_{\alpha,a,b}$ is a lifting of a monotone degree one map $\varphi_{\alpha,a,b} : S^1 \to S^1$, which semiconjugates $f_{a,b}$ with the α-multiplication map $D_\alpha : x \mapsto \alpha x \pmod 1$.

Proof. Throughout this proof we will denote $\Phi_{\alpha,a,b} = \Phi_{a,b}$ and $\varphi_{\alpha,a,b} = \varphi_{a,b}$.

We start the proof by showing that $F_{a,b}$ has a fixed point. We will use property (2) from Proposition 2.2.1 and the fact that $F_{a,b}$ is continuous. Define $G_{a,b}(x) = F_{a,b}(x) - x$, clearly $G_{a,b}$ is continuous. We have that $G_{a,b}(0) = F_{a,b}(0) - 0 = a$. Now consider

$$G_{a,b}(-\lceil|a|\rceil) = F_{a,b}(-\lceil|a|\rceil) + \lceil|a|\rceil = a - (\alpha - 1)\lceil|a|\rceil.$$

Now, for $\alpha \geq 2$ this gives $a - (\alpha - 1)\lceil|a|\rceil \leq 0$, so, there exists $\theta_1 \in \mathbb{R}$ such that $G_{a,b}(\theta_1) \leq 0$. We now consider,

$$G_{a,b}(\lceil|a|\rceil) = F_{a,b}(\lceil|a|\rceil) - \lceil|a|\rceil = a + (\alpha - 1)\lceil|a|\rceil.$$

Hence for $\alpha \geq 2$, we have $a + (\alpha - 1)\lceil|a|\rceil \geq 0$, thus, for some $\theta_2 \in \mathbb{R}$, we have $G_{a,b}(\theta_2) \geq 0$.

Hence by the intermediate value theorem, as $G_{a,b}$ is continuous there exists $\theta_1 \leq x_{a,b} \leq \theta_2$ such that $G_{a,b}(x_{a,b}) = 0$, which means that $G_{a,b}(x_{a,b}) = F_{a,b}(x_{a,b}) - x_{a,b} = 0$. Therefore there exists $x_{a,b} \in \mathbb{R}$ such that $F_{a,b}(x_{a,b}) = x_{a,b}$, i.e. $F_{a,b}$ has a fixed point $x_{a,b}$.

By property (2) of Proposition 2.2.1 we have that $F_{a,b}(x_{a,b} + k) = x_{a,b} + \alpha k$. We now prove by induction on n that

$$F_{a,b}^n(x_{a,b} + k) = x_{a,b} + \alpha^n k \qquad (2.22)$$

for any $k \in \mathbb{Z}$ and $n \geq 0$.

For $n = 0$ we have

$$F_{a,b}^0(x_{a,b} + k) = x_{a,b} + k = x_{a,b} + \alpha^0 k,$$

so (2.22) holds in this case.

We assume now that (2.22) holds for $n = m$ and we want to prove it holds for $n = m + 1$. We have

$$F_{a,b}^{m+1}(x_{a,b} + k) = F_{a,b}(F_{a,b}^m(x_{a,b} + k)) = F_{a,b}(x_{a,b} + \alpha^m k) = x_{a,b} + \alpha^{m+1} k.$$

Hence we have for any $k \in \mathbb{Z}$ and $n \geq 0$, $F_{a,b}^n(x_{a,b} + k) = x_{a,b} + \alpha^n k$.

For any $n \geq 0$ we get $x_{a,b} + \alpha^n k \leq F_{a,b}^{m+n}(x) \leq x_{a,b} + \alpha^n(k + 1)$. Therefore, dividing by α^{m+n} and after simplifying, we have

$$\frac{x_{a,b}}{\alpha^{m+n}} + \frac{k}{\alpha^m} \leq \frac{F_{a,b}^{m+n}(x)}{\alpha^{m+n}} \leq \frac{x_{a,b}}{\alpha^{m+n}} + \frac{k+1}{\alpha^m}.$$

This implies that for any $r, s \geq n$ and every x we get

$$\left| \frac{F_{a,b}^r(x)}{\alpha^r} - \frac{F_{a,b}^s(x)}{\alpha^s} \right| \leq \frac{1}{\alpha^m}. \tag{2.23}$$

Hence the sequence $\left(\frac{F_{a,b}^r(x)}{\alpha^r} \right)_{r=1}^{\infty}$ satisfies the uniform Cauchy's condition so it converges uniformly.

As (2.21) converges uniformly and $\frac{F_{a,b}^n(x)}{\alpha^n}$ is continuous and increasing, we have that $\Phi_{a,b}$ is continuous and increasing.

We now prove that $\Phi_{a,b}(x + k) = \Phi_{a,b}(x) + k$ for any $k \in \mathbb{Z}$. We have that

$$\Phi_{a,b}(x + k) = \lim_{n \to \infty} \frac{F_{a,b}^n(x + k)}{\alpha^n} = \lim_{n \to \infty} \frac{F_{a,b}^n(x) + \alpha^n k}{\alpha^n} = \Phi_{a,b}(x) + k.$$

It remains to show that $\Phi_{a,b}(F_{a,b}(x)) = \alpha \Phi_{a,b}(x)$. We have

$$\Phi_{a,b}(F_{a,b}(x)) = \lim_{n \to \infty} \frac{F_{a,b}^n(F_{a,b}(x))}{\alpha^n} = \alpha \lim_{n \to \infty} \frac{F_{a,b}^{n+1}(x)}{\alpha^{n+1}} = \alpha \Phi_{a,b}(x).$$

The final statement of this lemma follows directly from this as we have shown that $\Phi_{a,b}$ satisfies the definition of a lift. □

In the following lemmas, we continue to build up machinery similar to that in [27].

Lemma 2.2.3. *If p is a periodic point of $f_{\alpha,a,b}$ of period n then $\varphi_{\alpha,a,b}(p)$ is a periodic point of D_α of period n.*

Proof. Throughout this proof we denote $f_{\alpha,a,b} = f_{a,b}$ and also $\varphi_{\alpha,a,b} = \varphi_{a,b}$. Recall that $f_{\alpha,a,b}$ satisfies properties (1) and (2) of Proposition 2.2.1.

Assume p is a periodic point of $f_{a,b}$ of period n. We know that $\varphi_{a,b}(x)$ semiconjugates $f_{a,b}$ with D_α, i.e. $\varphi_{a,b}(f_{a,b}(x)) = D_\alpha(\varphi_{a,b}(x))$. We use this identity n times when $x = p$ and the fact that $f_{a,b}^n(p) = p$ and we get

$$\varphi_{a,b}(p) = \varphi_{a,b}(f_{a,b}^n(p)) = \alpha^{n-1}\varphi_{a,b}(f_{a,b}(p)) = \alpha^n\varphi_{a,b}(p) = D_\alpha^n(\varphi_{a,b}(p)).$$

So $\varphi_{a,b}(p)$ is a periodic point of D_α. It remains to show that it has period n. Assume for a contradiction it has period different from n, then by the fact that $D_\alpha^n(\varphi_{a,b}(p)) = \varphi_{a,b}(p)$ it must have period dividing n. Say the period of $\varphi_{a,b}$ under D_α is m and let η_1 be a point on the orbit of p so for some $1 \le e \le n$, $f_{a,b}^e(p) = \eta_1$. Now let η_2 be another point on the orbit of p such that $f_{a,b}^{e+m}(p) = \eta_2$ clearly η_1 and η_2 are distinct, it follows that,

$$\varphi_{a,b}(\eta_2) = \alpha^{e+m}\varphi_{a,b}(p) = \alpha^e\varphi_{a,b}(p) = \varphi_{a,b}(\eta_1),$$

and so there are points $x \ne y$ on the orbit of p which are mapped to the same point under $\varphi_{a,b}$. In Lemma 2.2.2 we established the monotonicity of $\varphi_{a,b}$. If we construct an arc A joining x to y we will show this arc is mapped by $\varphi_{a,b}$ to one point. If we assume that there exists c on our arc A such that $x < c < y$ if $\varphi_{a,b}(c) < \varphi_{a,b}(x)$ then $\varphi_{a,b}(c) < \varphi_{a,b}(y)$ contradicting monotonicity of $\varphi_{a,b}$. Similarly if $\varphi_{a,b}(c) > \varphi_{a,b}(x)$ then $\varphi_{a,b}(c) > \varphi_{a,b}(y)$, again contradicting monotonicity. So $\varphi_{a,b}$ maps all points on A to a single point. We can assume the arc A can be taken in the anticlockwise direction. Let q be any point on the orbit of p, so there exists a k such that $f_{a,b}^k(x) = q$ where x is as above. Then we can see that $f_{a,b}^k(A)$ describes an arc joining q to the point on the orbit equal to $f_{a,b}^{k+m}(p)$. In particular $f_{a,b}^k(A)$ contains an arc joining q with its anticlockwise neighbour on the orbit of p and as we have established is mapped by $\varphi_{a,b}$ to one point. This means the whole circle is mapped by $\varphi_{a,b}$ to one point which is a contradiction. Hence we must have that the period of $\varphi_{a,b}$ under D_α must be n. □

For the family of double standard maps in [27] (see (1.1)) we can easily show that its boundary is given by the curves

$$a = \frac{1}{2} \pm \frac{\sqrt{4b^2 - 1} - \arctan\sqrt{4b^1 - 1}}{2\pi} \qquad (2.24)$$

and the corresponding fixed point is

$$x = -\frac{1}{2} \pm \frac{\arctan\sqrt{4b^1 - 1}}{2\pi}. \qquad (2.25)$$

Let us now see what happens for α-standard maps.

Example 2.2.4. Let us show that for the family of maps (1.3), the boundary of the Arnold tongue of period 1 is given by,

$$a = (\alpha - 1)\left(\frac{1}{2} \pm \frac{\sqrt{\frac{\alpha^2}{(\alpha-1)^2}b^2 - 1} - \arctan\left(\sqrt{\frac{\alpha^2}{(\alpha-1)^2}b^2 - 1}\right)}{2\pi}\right), \qquad (2.26)$$

and the corresponding fixed point is

$$x = -\frac{1}{2} \pm \frac{\arctan\left(\sqrt{\frac{\alpha^2}{(\alpha-1)^2}b^2 - 1}\right)}{2\pi}.$$

We have $F'_{a,b}(x) = \alpha(1 + b\cos(2\pi x))$. Thus, solving $F'_{a,b}(x) = 1$ we get

$$\cos(2\pi x) = \frac{1}{b}\left(\frac{1}{\alpha} - 1\right).$$

Note that $\frac{1}{\alpha} - 1 \leq 0$ as $\alpha \geq 1$, so as $b \geq 0$, we have that $\cos(2\pi x) \leq 0$, hence we have

$$2\pi x = -\pi \pm \arccos\left(\frac{1}{b}\left(1 - \frac{1}{\alpha}\right)\right).$$

Using the identity $\arccos(x) = \arctan\left(\frac{\sqrt{1-x^2}}{x}\right)$, valid for $x \geq 0$, we can rewrite $\arccos\left(\frac{1}{b}\left(1 - \frac{1}{\alpha}\right)\right)$ as

$$\arccos\left(\frac{1}{b}\left(1 - \frac{1}{\alpha}\right)\right) = \arctan\left(\sqrt{\frac{\alpha^2}{(\alpha-1)^2}b^2 - 1}\right).$$

Thus, we get

$$2\pi x = -\pi \pm \arctan\left(\sqrt{\frac{\alpha^2}{(\alpha-1)^2}b^2 - 1}\right),$$

and so,

$$x = -\frac{1}{2} \pm \frac{\arctan\left(\sqrt{\frac{\alpha^2}{(\alpha-1)^2}b^2 - 1}\right)}{2\pi}. \tag{2.27}$$

It remains to find the corresponding value of a. We find this by substituting the value of x we have found into $F_{a,b}(x) = x$ and rearranging for a. First we compute $\sin(2\pi x)$ on the boundary,

$$\sin(2\pi x) = \sin\left(2\pi\left(-\frac{1}{2} \pm \frac{\arctan\left(\sqrt{\frac{\alpha^2}{(\alpha+1)^2}b^2 - 1}\right)}{2\pi}\right)\right),$$

$$= -\sin\left(\pm\arctan\left(\sqrt{\frac{\alpha^2}{(\alpha+1)^2}b^2 - 1}\right)\right).$$

We can now use the identity $\sin(\pm\arctan x) = \pm\frac{x}{\sqrt{1+x^2}}$ to simplify our expression for $\sin(2\pi x)$.

$$\sin(2\pi x) = \mp\frac{(\alpha - 1)}{\alpha b}\sqrt{\frac{\alpha^2}{(\alpha - 1)^2}b^2 - 1}.$$

We substitute this into $F_{a,b}(x) = x$ and we get,

$$\alpha x + a + \frac{\alpha b}{2\pi}\left(\mp\frac{(\alpha - 1)}{\alpha b}\sqrt{\frac{\alpha^2}{(\alpha - 1)^2}b^2 - 1}\right) = x,$$

which gives that

$$a = -x(\alpha - 1) \pm \frac{(\alpha - 1)}{2\pi}\sqrt{\frac{\alpha^2}{(\alpha - 1)^2}b^2 - 1}.$$

We now simplify this result and use our value of x given by (2.27) to see that,

$$a = (\alpha - 1)\left(\frac{1}{2} \pm \frac{\sqrt{\frac{\alpha^2}{(\alpha-1)^2}b^2 - 1} - \arctan\left(\sqrt{\frac{\alpha^2}{(\alpha-1)^2}b^2 - 1}\right)}{2\pi}\right).$$

In Figures 2.8–2.11, we plot the Arnold tongues with a taken modulo 1. The plots show a along the x-axis from 0 to 1 and b along the y-axis from 0 to 1.

Note that for the Arnold tongues if α is even then the centre point of the tongue of period 1 is at $a = 0.5$,

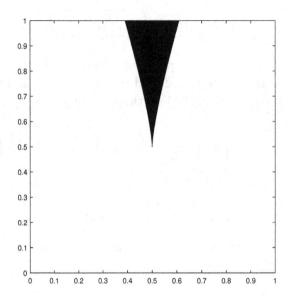

Figure 2.8: Arnold tongue of period 1, when $\alpha = 2$.

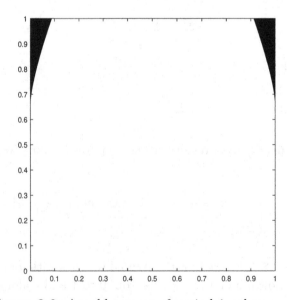

Figure 2.9: Arnold tongue of period 1, when $\alpha = 3$.

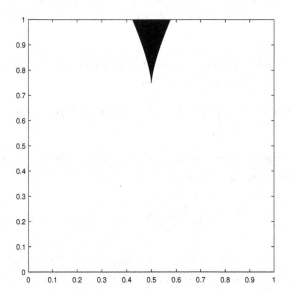

Figure 2.10: Arnold tongue of period 1, when $\alpha = 4$.

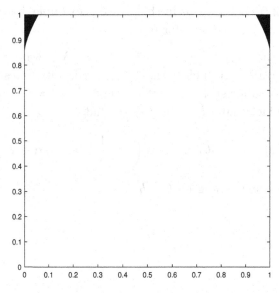

Figure 2.11: Arnold tongue of period 1, when $\alpha = 7$.

if α is odd the centre point is at $a = 0$. We also notice that as α increases the value of b at the bottom or tip of the tongue increases.

We would like to study the rate at which the height of the tip changes, the next example gives us this result.

Example 2.2.5. Denote the tip of the period 1 tongue by (a_0, b_0). Let us show that $b_0 = 1 - \frac{1}{\alpha}$.

Let b_0 be the smallest value of b for which we have real values in our expression for a and x, this is when we have

$$\frac{\alpha^2}{(\alpha - 1)^2} b_0^2 - 1 = 0.$$

Solving for b_0 we get $b_0 = \pm \frac{\alpha - 1}{\alpha}$. Taking the positive solution as $b > 0$ we see that the tip of the tongue occurs at

$$b_0 = 1 - \frac{1}{\alpha}.$$

Example 2.2.6. Let b_0 be as in the previous example. Let us show that the map $f_{a,b}(x)$ is expanding for $0 \le b < b_0$.

We have that $f'_{a,b}(x) = \alpha + \alpha b \cos(2\pi x)$. We want to show that $f'_{a,b}(x) > 1$ for all x. It is clear that $f'_{a,b}(x)$ takes its lowest value when $\cos(2\pi x) = -1$ so for such an x we see that $f'_{a,b}(x) = \alpha - \alpha b$.
Using the fact that $-b > -b_0$ we see that

$$f'_{a,b}(x) > \alpha - \alpha b_0 > \alpha - \alpha \left(1 - \frac{1}{\alpha}\right) > \alpha - \alpha + 1 > 1.$$

Hence $f_{a,b}(x)$ is expanding for $0 \le b < b_0$.

CHAPTER 3

Topological conjugacy

We define topological conjugacy as follows.

Definition 3.0.1. Given two maps $f : A \to A$ and $g : B \to B$ we say they are *topologically conjugate* if there exists a homeomorphism $\pi : A \to B$ such that $\pi \circ f = g \circ \pi$. We (sometimes) write $f \sim g$ meaning that f and g are *topologically conjugate*.

$$
\begin{array}{ccccccc}
a_0 & \xrightarrow{f} & a_1 & \xrightarrow{f} & \cdots & \xrightarrow{f} & a_n \\
\downarrow{\pi} & & \downarrow{\pi} & & & & \downarrow{\pi} \\
b_0 & \xrightarrow{g} & b_1 & \xrightarrow{g} & \cdots & \xrightarrow{g} & b_n
\end{array}
$$

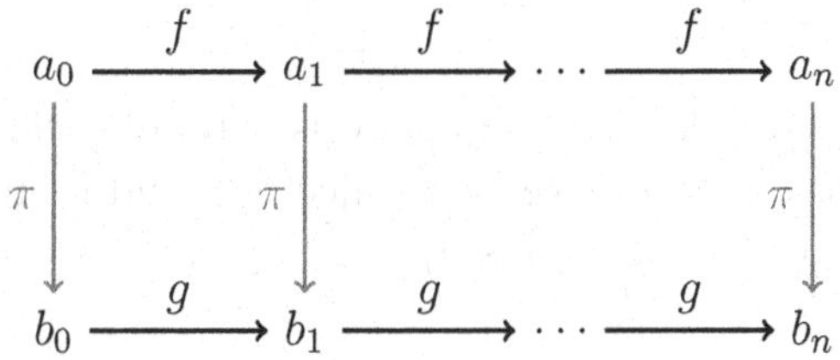

Figure 3.1: Representation of a topological conjugacy. We show the orbit of a_0 under f with $f(a_i) = a_{i+1}$ for $i = 0, \ldots, n$ in A and similarly, the orbit of b_0 under g with $g(b_i) = b_{i+1}$ for $i = 0, \ldots, n$ in B. We say f and g are topologically conjugate via π.

Example 3.0.2. Consider for example in $[0,1]$ the tent map T, and the quadratic map F_μ:

$$
T(y) = \begin{cases} 2y & 0 \leq z \leq \frac{1}{2} \\ 2(1-y) & \frac{1}{2} \leq z \leq 1 \end{cases}
$$

and

$$
F_\mu = \mu z(1-z).
$$

Let us show that T and F_4 are topologically conjugate via the function $\sigma : [0, 1] \to [0, 1]$, where $\sigma(z) = \sin^2(\pi z)$. We have

$$F_4 \circ \sigma(z) = F_4(\sigma(z)) = 2\sin^2(\pi z) - 4\sin^4(\pi z)$$
$$= 4\sin^2(\pi z)\cos^2(\pi z) = (2\sin(\pi z)\cos(\pi z))^2$$
$$= \sin^2(2\pi x).$$

Now for the tent map, when $0 \le z \le 1/2$, we have

$$\sigma \circ T(z) = \sigma(T(z)) = \sin^2(2\pi z),$$

and when $1/2 \le z \le 1$, we get

$$\sigma \circ T(z) = \sigma(T(z)) = \sin^2(2\pi - 2\pi z) = \sin^2(2\pi z).$$

So as required, we have $\sigma \circ T = F_4 \circ \sigma$. Therefore T and F_4 are topologically conjugate via σ.

Definition 3.0.3. Given two maps $f : A \to A$ and $g : B \to B$, we say these maps are *topologically semiconjugate* if there exists a surjection φ such that $f \circ \varphi = \varphi \circ g$.

In a sense, every topological conjugacy is a topological semiconjugacy, just the map φ is also injective, so a bijection, and its inverse is continuous (making it a homeomorphism).

Example 3.0.4. Consider now the doubling map D and again the tent map:

$$D(x) = 2x \mod 2,$$

and

$$T(y) = \begin{cases} 2y & 0 \le y \le \frac{1}{2}, \\ 2(1-y) & \frac{1}{2} \le y \le 1. \end{cases}$$

We now show that these two systems are topologically semi-conjugate. Let $\varphi : [0, 2] \to [0, 1]$ be defined by:

$$\varphi(x) = \begin{cases} x & 0 \le x \le 1, \\ 2 - x & 1 \le x \le 2. \end{cases}$$

Then when $0 \le x \le 1/2$, $T \circ \varphi(x) = T(\varphi(x)) = T(x) = 2x$ and $\varphi \circ D = \varphi(D(x)) = \varphi(2x) = 2x$ because $0 \le 2x \le 1$.

Now if $1/2 \leq x \leq 1$, then $T \circ \varphi = T(\varphi(x)) = T(x) = 2 - 2x$, and $\varphi \circ D = \varphi(D(x)) = \varphi(2x) = 2 - 2x$ because $1 \leq 2x \leq 2$.

So get that $T \circ \varphi = \varphi \circ D$. This is only a semiconjugacy, as we have a two-to-one correspondence from the space of D to the space of T.

When we are modelling dynamical systems, it is often important to consider the stability of the system under small changes or perturbations.

Definition 3.0.5. If every map that is classified to be *nearby* a given map f, is topologically conjugate to f, then f is said to be *structurally stable*.

This is important in applications of dynamical systems, normally the model will be a model of a real-world phenomena. If a system is inherently unstable, it will be heavily affected by any assumptions, approximations and experimental errors that will no doubt be present in the model. If we can be assured a system is stable, then it is more likely to produce a more accurate representation of the data being modelled.

Example 3.0.6. As another example, let us now consider the quadratic map, $F_{4.5}$, the map $S(x) : \Lambda \to \Sigma_2$, and the shift map σ. We can show that the shift map and the quadratic map are conjugates via $S(x)$: $S \circ F_{4.5} = \sigma \circ S$. To see this, recall that $x \in \Lambda$ may be represented by the nested intervals: $\bigcap_{n \geq 0} I_{s_0, s_1, \ldots}$ which is given by the itinerary: $S(x)$. We can write

$$I_{s_0, \ldots s_n} = I_{s_0} \cap F_{4.5}^{-1}(I_{s_1}) \cap \cdots \cap F_{4.5}^{-n}(I_{s_n}),$$

since $F_{4.5}(I_0) = I$. And so we have

$$SF_{4.5}(x) = SF(\bigcap_{n=0}^{\infty} I_{s_0 \ldots s_n}) = S(\bigcap_{n=1}^{\infty} I_{s_1 \ldots s_n}) = s_1 s_2 \cdots = \sigma S(x).$$

Hence the dynamics of the shift map on Σ_2 is equivalent to the dynamics of the quadratic map $F_{4.5}$ on Λ.

In [27] it was proved that our family of double standard maps given by

$$f_{a,b}(x) = 2x + a + \frac{b}{\pi} \sin(2\pi x) \mod 1,$$

is topologically conjugate to

$$g_{a,b}(z) = e^{2\pi a i} z^2 e^{b(z - \frac{1}{z})}$$

via $\psi = e^{2\pi x i}$ for $z \in \mathbb{C}$ on the unit circle.

Let us examine our main example in more detail. Clearly, we have for $0 \le x \le 1$ that

$$\psi \circ f_{a,b} = e^{(2\pi i)(2x + a + \frac{b}{\pi}\sin(2\pi x))}$$

$$= e^{2\pi a i} e^{(2\pi i)(2x)} e^{2bi \sin(2\pi x)}$$

$$= e^{2\pi a i} \left(e^{2\pi i x}\right)^2 e^{2bi \sin(2\pi x)}.$$

Also

$$g_{a,b} \circ \psi = e^{2\pi a i} \left(e^{2\pi i x}\right)^2 e^{b(e^{2\pi x i} - e^{-2\pi x i})}$$

$$= e^{2\pi a i} \left(e^{2\pi i x}\right)^2 e^{2bi \sin(2\pi x)}.$$

From this we get

$$\psi \circ f_{a,b} = g_{a,b} \circ \psi,$$

hence $f_{a,b}$ and $g_{a,b}$ are topologically conjugate.

The existence of this topological conjugacy allowed us to prove the following theorem (as in [27]) using techniques from Complex Dynamics.

Theorem 3.0.7. *If $0 \le b \le 1$ then $f_{a,b}$ has at most one attracting or neutral periodic orbit.*

The proof of this theorem involved conjugacy of $f_{a,b}$ via $e^{2\pi i x}$ with the complex map

$$g_{a,b}(z) = e^{2\pi i a} z^2 e^{b\left(z - \frac{1}{z}\right)}, \tag{3.1}$$

of the unit circle to itself. For this map, any immediate basin of attraction of an attracting or neutral periodic orbit has to contain a critical point, and there is only one pair of critical points, symmetric (in the complex sense) with respect to the unit circle. The map also preserves this symmetry and the theorem follows.

CHAPTER 4

Critical points

Recall our family of double standard maps in 1.1

$$f_{a,b}(x) = 2x + a + \frac{b}{\pi}\sin(2\pi x) \ (\text{mod } 1).$$

Let us consider the behaviour of the critical points for values of $b \in [0,1]$. Clearly,

$$\frac{d}{dx}f_{a,b}(x) = 2 + 2b\cos(2\pi x).$$

When $b = 1$ there is one critical point, where $\cos(2\pi x) = -1$. However, when $0 < b < 1$ there is a pair of complex conjugate critical points given by

$$x = \frac{1}{2\pi}\cos^{-1}\left(-\frac{1}{b}\right).$$

Theorem 4.0.1. *These complex points all lie on the line with real value* $\frac{1}{2}$*, and as* $b \to 1$*, these points converge to the value* $x = \frac{1}{2}$*.*

Proof. We can write the derivative of $f(z)_{a,b}$ in terms of exponents, and then separate $z = x + iy$ into real and complex parts ($a, b \in \mathbb{R}$), as we know we are looking for complex solutions:

$$\frac{d}{dx}f_{a,b}(z) = 2 + be^{-2\pi iz} + be^{2\pi iz},$$

and we get

$$-2 = b\left(e^{-2\pi ix}e^{2\pi y} + e^{2\pi ix}e^{-2\pi y}\right).$$

Then we can express the complex exponentials in terms of cos and sin, and separate out real and complex parts of the equation:

$$-2 = b\left(e^{2\pi y}(\cos 2\pi x - i\sin 2\pi x) + e^{-2\pi y}(\cos 2\pi x + i\sin 2\pi x)\right).$$

Simplifying, we get

$$-2 = b\left(\cos 2\pi x\left(e^{2\pi y} + e^{-2\pi y}\right) + i\sin 2\pi x\left(e^{-2\pi y} - e^{2\pi y}\right)\right).$$

Next we note that we are looking for a purely real solution, so we need to lose the complex part of the equation above. Because $e^{-2\pi y} - e^{2piy} \neq 0$ $\forall y \in \mathbb{R} \setminus 0$, we need $\sin(2\pi x) = 0$. We choose to ignore the case $y = 0$ because then for $b < 1$ we would need $\cos 2\pi x > 1$ which is not possible for $x \in \mathbb{R}$. Therefore $x = n/2$ $n \in \mathbb{Z}$.

We can restrict x further by realising that $e^{2\pi y} + e^{-2\pi y} > 0 \forall y \in \mathbb{R}$, and so we also require $\cos 2\pi x < 0$, therefore $x = 3n/2, n \in \mathbb{Z}$. Now the only solution remaining in the domain $[0, 1]$ is $x = 1/2$. So we have proved that the real part remains constant at $1/2$.

For the converging complex conjugate part, we can see the equation now reduces to the following:

$$2 = b(e^{2\pi y} + e^{-2\pi y}) = 2b\cosh 2\pi y$$

cosh is symmetric about 0, so if y is a solution, then so is $-y$. The solution when $b = 1$ is $y = 0$, so because cosh is a smooth function, as $b \to^- 1$, the solutions $y \to 0$ from both the positive and negative side, proving the pair of converging conjugate critical points as $b \to^- 0$.

□

We will now explore how the orbit and periodicity of the critical point changes under $f_{a,b}$ when $b = 1$ is fixed, and a is varied. We have seen that the orbit remains fixed at $x = 1/2$ when $a = 0.5$, so let us investigate what happens when $a = 0.6, 0.7, 0.8, 0.9$.

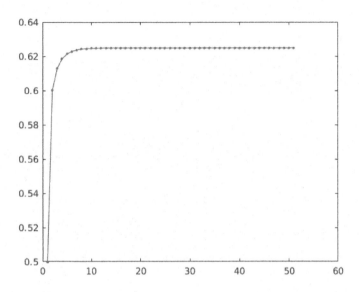

Figure 4.1: The orbit of the critical point $x_0 = 1/2$ under $f_{0.6,1}$. A periodic orbit of period 1 can be seen.

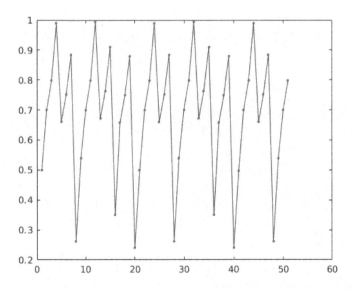

Figure 4.2: The orbit of the critical point $x_0 = 1/2$ under $f_{0.7,1}$, a periodic orbit of period 20 can be seen.

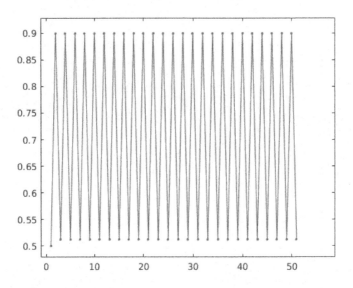

Figure 4.3: The orbit of the critical point $x_0 = 1/2$ under $f_{0.9,1}$. A periodic orbit of period 2 can be seen.

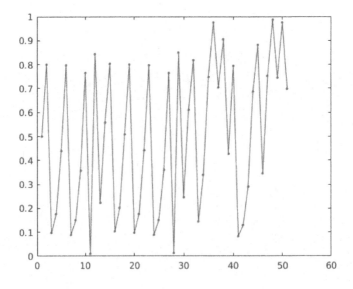

Figure 4.4: The orbit of the critical point $x_0 = 1/2$ under $f_{0.8,1}$. Periodic behaviour is not apparent.

It is apparent then, that under these conditions the behaviour of $f_{a,b}$

is wildly different for different values of a. We can see that the value $a = 0.6$ lies in the middle region for where there is only one fixed point, just as Figure 4.1 suggests. We can also see the region of period two containing $a = 0.9$ that we expect from Figure 4.3.

The chaotic behaviour in Figure 4.4 is supported by Figure 4.5 in that the a value of 0.8 lies in a region that is very dense, hence we do not expect to observe any periodic behaviour. And while we observed a period of 20 for the a value 0.7, that is difficult to discern from Figure 4.5 because the period is very large, and the region for which this periodic behaviour occurs is very small.

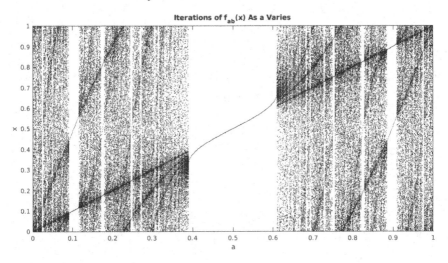

Figure 4.5: A plot of the most common points in the iteration of $x_0 = 0.5$ under $f_{a,1}$ as a varies in $0 \leq a \leq 1$.

In Figure 4.5, there is a form of symmetry in the arising structure. The dynamics in the region $0.5 < a < 1$ mirror that of the region $0 < a < 0.5$, in the lines $a = 0.5$ and $x = 0.5$. In order to understand why this structure arises, let us now consider the orbits of critical points and conjugate points under $g_{a,b}$.

In Figures 4.6—4.9 we take the orbits of three points under the iteration of $g_{a,b}$. The point on each plot in question is given by a red star, but explicitly the left plot takes the orbit of $x = i$, and the right takes the conjugate point $x = -i$. The middle plot expresses the orbit of the critical point when $b = 1$, namely $x = -1$.

Note that the orbit of $\bar{x}$ in the right of Figure 4.7 is the reflection in the real axis of the x on the left of Figure 4.6, and vice versa for the orbit of $\bar{x}$ in Figure 4.6 and x in Figure 4.7. Similarly, the reflection of the orbit of the critical point in the x axis in Figure 4.7 is the same as the orbit of the critical point in Figure 4.6. We can see the same relationship holds between Figures 4.8 and 4.9. This is a result of the symmetry we noticed in Figure 4.5.

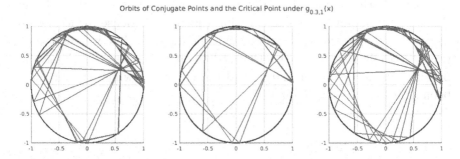

Figure 4.6: Orbit of conjugate points x and $\bar{x}$ for $x = e^{1.2\pi i}$ under the first 100 iterations of $g_{0.3,1}$.

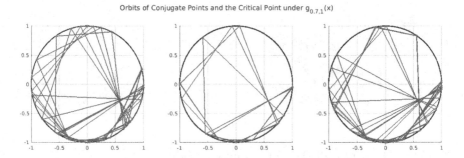

Figure 4.7: Orbit of conjugate points x and $\bar{x}$ for $x = e^{1.2\pi i}$ under the first 100 iterations of $g_{0.7,1}$.

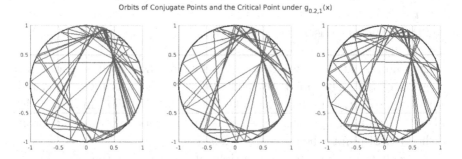

Figure 4.8: Orbit of conjugate points x and $\bar{x}$ for $x = e^{1.2\pi i}$ under the first 100 iterations of $g_{0.2,1}$.

Orbits of Conjugate Points and the Critical Point under $g_{0.8,1}(x)$

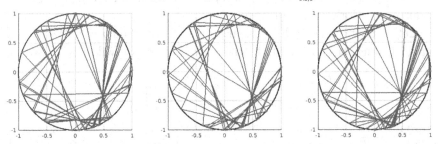

Figure 4.9: Orbit of conjugate points x and $\bar{x}$ for $x = e^{1.2\pi i}$ under the first 100 iterations of $g_{0.8,1}$.

What could explain this relation? Looking at the a values used for these orbits, we notice that this observed relationship between two orbits occurs when the a values of the corresponding functions are given by a and $1 - a$ respectively. We have the following lemma.

Lemma 4.0.2. *Recall our family of maps given by* (1.2)

$$g_{a,b}(z) = e^{2\pi i a} z^2 e^{b\left(z - \frac{1}{z}\right)}. \tag{4.1}$$

The following holds

$$\overline{g_{a,b}(x)} = g_{1-a,b}(\bar{x}), \tag{4.2}$$

where $\bar{x}$ denotes the complex conjugate of x.

Proof. We have

$$g_{a,b}(x) = e^{2\pi a i} x^2 e^{b(x - \frac{1}{x})} \Rightarrow \overline{g_{a,b}(x)} = \overline{e^{2\pi a i} x^2 e^{b(x - \frac{1}{x})}}.$$

Thus

$$\overline{g_{a,b}(x)} = \left(\overline{e^{2\pi a i}}\right)\left(\overline{x^2}\right)\left(\overline{e^{b(z - \frac{1}{z})}}\right) = e^{-2\pi a i} \bar{x}^2 e^{b(\bar{x} - \frac{1}{\bar{x}})}.$$

Now given that $e^{\pi i} = 1$, $e^{2\pi i} = 1^2 = 1$, we can multiply the first term on the right hand side by $e^{2\pi i}$ and we get

$$\overline{g_{a,b}(x)} = e^{2\pi i (1-a)} \bar{x}^2 e^{b(\bar{x} - \frac{1}{\bar{x}})} = g_{(1-a),b}(\bar{x})$$

and we finished the proof that $\overline{g_{a,b}(x)} = g_{1-a,b}(\bar{x})$.

$\square$

There exists a kind of symmetry in the orbits of conjugate points under $g_{a,b}(x)$ for mirrored a values.

Let us now consider the critical points of $g_{a,b}$ when $b = 1$. We should expect to find similar results to those for $f_{a.b}$ but we will clarify this:

$$\frac{d}{dx}g_{a,b}(x) = e^{2\pi ai + b(x - \frac{1}{x})}(bx^2 + 2x + b).$$

Implying critical points occur when $bx^2 + 2x + b = 0$, so $x = 1/b(\pm\sqrt{1 - b^2} - 1)$. When $b = 1$ $x = -1$ Otherwise for $0 < b < 1$ there does not exist a solution on the unit circle, which is the domain we are concerning ourselves with. Looking at Figure 4.10 we can see this graphically: when $b = 1$ we have a critical point at $arg(x) = \pi$ which corresponds to $x = -1$ on the unit circle as we expected. One can also see the fixed points of $g_{a,b}$ here too where the grey diagonal meets the line of $g_{a,b}$ (note this plot is a little misleading, as the fixed points are only where the lines cross where there are not discontinuities, so in general, the vertical lines do not imply a fixed point).

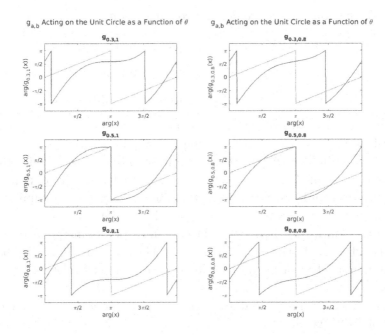

Figure 4.10: Plot of $g_{a,b}(x)$ as a function of θ, the argument of x on the unit circle. Critical points can be seen clearly in all three plots on the left-hand side where $b = 1$ ($\theta = \pi$), otherwise there are no critical points when $0 < b < 1$. The diagonal dotted lines mark the identity mod 2π, points where the two plots cross mark fixed points on the unit circle (only when neither of the lines occur at a discontinuity).

We are interested in explaining some more of the features of Figure 4.5, so we are only concerned in cases where $b = 1$. To begin we notice from Figure 4.10 that while $g_{a,b}$ has 3 fixed points when $a = 0.5$, it only features one for $a = 0.3, 0.8$. This suggests there exists some form of bifurcation point for fixed points in the regions $[0.3, 0.5]$ and $[0.5, 0.8]$.

Consider the equation for the fixed points of $f_{a,b}(x)$:

$$a = \frac{1}{2} \pm \frac{\sqrt{4b^2 - 1} - \arctan\sqrt{4b^2 - 1}}{2\pi}. \tag{4.3}$$

By taking $b < 0.5$, we should expect there to be no values for a in which bifurcation points exist. And indeed, when $b = 0.5$, $\sqrt{4b^2 - 1} - \arctan\sqrt{4b^2 - 1} = 0 - 0 = 0$, and so there is one solution for a, $1/2$, and when $b < 1/2$ there are no real solutions for a whatsoever.

With this knowledge, we define two values of a,

$$a^+ = 1/2 + (\sqrt{3} - \arctan\sqrt{3})/(2\pi) \approx 0.61$$

and

$$a^- = 1 - a^+ = 1/2 - (\sqrt{3} - \arctan\sqrt{3})/(2\pi) \approx 0.39,$$

which are the two solutions to equation (4.3) for $b = 1$. These are the values for which $a \in [a^+, a^-]$ implies $g_{a,1}$ has 3 fixed points, and $a \in [a^+, a^-]^c$ implies $g_{a,1}$ has only one fixed point. We can infer from our knowledge of $f_{a,b}$ that when $a = 1/2$, we have a fixed point at $x = e^{i\pi} = -1$ and two repelling fixed points. We know the number of fixed points must remain the same for a in the interval (a^-, a^+), and Figure 4.11 illustrates how the rightmost fixed points merge at $a = a^+$ and the leftmost do the same for $a = a^-$. Then for $a \in (0, a^+)$ we have only a unique leftmost fixed point, as seen in Figure 4.10, let us call this left point $p(a)$.

Now since we know that $g_{a,1}(p(a)) = p(a)$ for $a \in (a^+, 1)$, recall equation (4.2) which tells us

$$g_{1-a,1}\overline{[p(a)]} = \overline{p(a)}.$$

Then the conjugate of $p(a)$ for $a \in (a^+, 1)$ is the unique fixed point for a corresponding $a \in (0, a^-)$ due to the fact $a^+ = 1 - a^-$. So for $a \in [0, a^-]$ the system is symmetric.

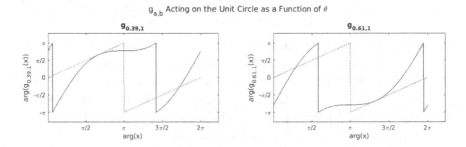

Figure 4.11: Plot of $g_{a,1}(x)$ as a function of θ, the argument of x on the unit circle. Here we have chosen [LEFT] $a = a^-$ and [RIGHT] $a = a^+$. The merging of fixed points can be seen.

CHAPTER 5

Topological theory of chaos

The use of the term chaos was introduced in Dynamical Systems by Li and Yorke [22] (if a map on the line has a periodic point of period three then it has points of all periods). A map f on a metric space X is chaotic on an invariant set Y provided f is transitive on.Y (if given two open sets U and V in Y there exists a positive integer n such that $f^n(U) \cap V \neq \emptyset$ and f has sensitive dependence on initial conditions on Y (if there exists $r > 0$ such that for each $x \in X$ and for each $\epsilon > 0$ there is $y \in X$ with $d(x, y) < \epsilon$ and $k \leq 0$ such that $d(f^k(x), f^k(y)) \geq r$.

We can now describe the structure of the periodic points of a map f.

Proposition 5.0.1. *If f^k is chaotic for some $k \geq 1$, then f is also chaotic.*

Consider the following order (*Sarkovskii ordering*):

$$3, 5, 7, \ldots, 2 \times 3, 2 \times 5, 2 \times 7, \ldots, 2^2 \times 3, 2^2 \times 5, 2^2 \times 7, \ldots, 2^3, 2^2, 1.$$

Theorem 5.0.2. *(Sarkovsii Theorem). Suppose that f has a periodic point of period k. If n stands to the right of k in the Sarkovsii ordering then f has a periodic point of period n.*

From this theorem we have a sufficient condition for the existence of chaos.

Theorem 5.0.3. *If f has a periodic point of a period which is not a power of 2 then f is chaotic.*

Proof. If we assume that f has a periodic point of a period which is not a power of 2 then we can write that $k = 2^p \times n$ for some p greater or equal to 0 and $n \neq 1$ odd.

The number $2^{p+1} \times 3$ stands to the right of k in the Sarkovskii ordering, hence by Sharkovskii Theorem, f has a periodic point of period $2^{p+1} \times 3$. Thus, $f^{2^{p+1}}$ has a periodic point of period 3, hence it is chaotic. Thus, f is also chaotic.

□

Let us state now a well-known theorem that tells us that *period 3 implies chaos.*

Theorem 5.0.4. *Let J be an interval and let $f : J \to J$ be continuous. Assume there is a point $a \in J$ for which the points $b = F(a), c = F^2(a)$ and $d = F^3(a)$, satisfy*

$$d \leq a < b < c$$

(or $d \geq a > b > c$). Then,

T1: for every $k = 1, 2, \dots$ there is a periodic point in J having period k.

Furthermore,

T2: there is an uncountable set $S \subset J$ (containing no periodic points), which satisfies the following conditions:

(A) For every $p, q \in S$ with $p \neq q$,

$$\limsup_{n \to \infty} |F^n(p) - F^n(q)| > 0,$$

and

$$\liminf_{n \to \infty} |F^n(p) - F^n(q)| = 0,$$

(B) For every $p \in S$ and periodic point $q \in J$,

$$\limsup_{n \to \infty} |F^n(p) - F^n(q)| > 0.$$

We will follow the proof of T1 in [22] we can also find the proof for T2). Let us first prove some lemmas that will be needed in order to prove T1.

Lemma 5.0.5. *Let $G : I \to \mathbf{R}$ be continuous, where I is an interval. For any compact interval $I_1 \subset G(I)$ there is a compact interval $Q \subset I$ such that $G(Q) = I$.*

Proof. Let $I_1 = [G(p), G(q)]$, where $p, q \in I$. If $p < q$ (the same proof is valid for $p > q$), let r be the last point of $[p, q]$, where $G(r) = G(p)$ and let s be the first point after r, where $G(s) = G(q)$. Then we have that $G([r, s]) = I_1$.

□

Lemma 5.0.6. *Let $f : J \to J$ be continuous and let $\{I_n\}_{n=0}^{\infty}$ be a sequence of compact intervals with $I_n \subset J$ and $I_{n+1} \subset F(I_n)$ for all n. Then there is a sequence of compact intervals Q_n such that $Q_{n+1} \subset Q_n \subset I_0$ and $F^n(Q_n) = I_n$ for $n \geq 0$. For any $x \in Q = \cap Q_n$, we have $F^n(x) \in I_n$ for all n.*

Proof. Define $Q_0 = I_0$. Then $F^n(Q_0) = I_0$. If Q_{n-1} has been defined so that $F^{n-1}(Q_{n-1}) = I_{n-1}$, then $I_n \subset F(I_{n-1}) = F^n(Q_{n-1})$. By Lemma 5.0.5 applied to $G = F^n$ on Q_{n-1} there is a compact interval $Q_n \subset Q_{n-1}$ such that $F^n(Q_n) = I_n$. This finishes the induction.

□

Lemma 5.0.7. *Let $G : J \to \mathbf{R}$ be continuous and let $I \subset J$ be a compact interval. Assume that $I \subset G(I)$. Then there is a point $p \in I$ such that $G(p) = p$.*

Proof. Let $I = [\beta_0, \beta_1]$. Choose α_i with $i = 0$ or $i = 1$ in I such that $G(\alpha_i) = \beta_i$. It follows that $\alpha_0 - G(\alpha_0) \geq 0$ and $\alpha_1 - G(\alpha_1) \leq 0$ and so continuity implies $G(\beta) - \beta$ must be 0 for some $\beta \in I$.

□

Proof of T1. We may now prove T1 following [22]. The proof of T2 can also be found in [22] but we will not state it here.

Assume $d \leq a < b < c$ as in our theorem (the proof for the other case is similar). Let $K = [a, b]$ and $L = [b, c]$ (again, we follow [22]).

Let k be a positive integer. For $k > 1$ let $\{I_n\}$ be the sequence of intervals $I_n = L$ for $n = 0, \ldots, k - 2$ and $I_{k-1} = K$.

Define I_n to be periodic inductively, this is, $I_{n+k} = I_n$ for $n = 0, 1, 2, \ldots$. If $k = 1$, let $I_n = L$ for all n.

Let Q_n be the sets in the proof of Lemma 5.0.6. Then $Q_k \subset Q_0$ and $F^k(Q_k) = Q_0$ and so by Lemma 5.0.7 $G = F^k$ has a fixed point $p_k in Q_k$. It is clear that p_k cannot have period less than k for F (otherwise we would need to have $F^{k-1}(p_k) = b$, contrary to $F^{k+1}(p_k) \in L$). The point p_k is a periodic point of period k for F. Thus, we finish our proof.

Note that we just proved that the existence of a point of period 3 implies the existence of one of period 5 however, the converse is false. A really nice example of this is again given in [22].

In their paper, Li and York present as an example the map $F :$ $[1,5] \rightarrow [1,5]$ such that

$$F(1) = 3, F(2) = 5, F(3) = 4, F(4) = 2, F(5) = 1$$

and on each interval $[n, n+1]$, with $1 \leq n \leq 4$ assume F is linear. Then we have

$$F([1,2]) = [3,5],$$
$$F([3,5]) = [1,4] = F^2([1,2]),$$
$$F([1,4]) = [2,3] = F^3([1,2]),$$
$$F([2,3]) = [4,5] = F^4([1,2]),$$
$$F([4,5]) = [1,2] = F^5([1,2]),$$

and thus, clearly F has an orbit of period 5.

5.1 TOPOLOGICAL ENTROPY

A mathematical tool to make the notion of chaos precise is topological entropy, a quantitative measurement of the amount of sensitive dependence of the map f on the initial conditions [31]. Topological entropy measures the rate of increase in dynamical complexity as a system evolves with time [34]. It was introduced as an invariant of topological conjugacy [1], which justifies its importance, as we know that mappings which are topologically conjugate are equivalent in terms of their dynamics. For instance, if we know that two maps are topologically conjugate, then we get correspondence between periodic points of both maps.

Topological entropy measures the increase in complexity as a system evolves with time. It was first introduced by [1], who defined it using open covers, but an equivalent definition was given by [7] (this paper was later updated in an erratum in [8]) and [17] by using spanning and separating sets.

Throughout this section, we will assume that X is a compact topological space. The following definitions were introduced by Adler, Konheim and McAndrew in [1].

Definition 5.1.1. Let $\mathcal{U}$ be an open cover of X. Denote the cardinality of the smallest subcover of $\mathcal{U}$ as $N(\mathcal{U})$, and define the *entropy* of the cover $\mathcal{U}$ to be

$$H(\mathcal{U}) = \log(N(\mathcal{U})).$$

Note that because X is compact, we have that the open cover $\mathcal{U}$ is guaranteed to have at least one finite subcover, and because the cardinality of a finite set is a positive integer, there must therefore be a subcover of the smallest cardinality.

Definition 5.1.2. For two open covers $\mathcal{U}$ and $\mathcal{V}$ of X, we define their join to be

$$\mathcal{U} \vee \mathcal{V} = \{U \cap V : U \in \mathcal{U}, V \in \mathcal{V}\}.$$

Definition 5.1.3. For two open covers $\mathcal{U}$ and $\mathcal{V}$ of X, we say that $\mathcal{V}$ is a *refinement* of $\mathcal{U}$, denoted $\mathcal{U} \prec \mathcal{V}$, if for all $V \in \mathcal{V}$, there is a $U \in \mathcal{U}$ such that $V \subseteq U$.

We have that $\vee$ is associative and commutative, and that $\prec$ is a reflexive partial order on the set of open covers of X, that is, $\prec$ is reflexive, antisymmetric and transitive. We write $f^{-1}(\mathcal{U})$ for the open cover $\{f^{-1}(U) : U \in \mathcal{U}\}$.

Lemma 5.1.4. *We have the following properties:*

1. *if $\mathcal{U} \prec \mathcal{V}$, then $H(\mathcal{U}) \leq H(\mathcal{V})$;*

2. $H(\mathcal{U} \vee \mathcal{V}) \leq H(\mathcal{U}) + H(\mathcal{V})$;

3. $f^{-1}(\mathcal{U} \vee \mathcal{V}) = f^{-1}(\mathcal{U}) \vee f^{-1}(\mathcal{V})$;

4. *if $\mathcal{U} \prec \mathcal{V}$, then $f^{-1}(\mathcal{U}) \prec f^{-1}(\mathcal{V})$;*

5. $H(f^{-1}(\mathcal{U})) \leq H(\mathcal{U})$.

Proof. Suppose $\mathcal{U} \prec \mathcal{V}$. Let $\{V_1, V_2, ..., V_{N(\mathcal{V})}\}$ be a subcover of $\mathcal{V}$. Then there is a subcover $\{U_1, U_2, ..., U_{N(\mathcal{V})}\}$ such that for each i, $V_i \subseteq U_i$. Therefore $N(\mathcal{U}) \leq N(\mathcal{V})$, and by the monotonicity of the logarithm, (1) is proved.

The join of the subcover of $\mathcal{U}$ with $N(\mathcal{U})$ elements and the subcover of $\mathcal{V}$ with $N(\mathcal{V})$ elements forms a cover consisting only of elements of the form $U_i \cap V_j$ for $1 \leq i \leq N(\mathcal{U}), 1 \leq j \leq N(\mathcal{V})$, where $U_i \in \mathcal{U}$ and $V_i \in \mathcal{V}$. Thus we have that $N(\mathcal{U} \vee \mathcal{V}) \leq N(\mathcal{U})N(\mathcal{V})$. This proves (2). Let $U \in \mathcal{U}$ and $V \in \mathcal{V}$. Then $f^{-1}(U \cap V) \in f^{-1}(\mathcal{U} \vee \mathcal{V})$ and

$$f^{-1}(U \cap V) = f^{-1}(U) \cap f^{-1}(V) \in f^{-1}(\mathcal{U}) \vee f^{-1}(\mathcal{V}),$$

proving (3).

We have that $\mathcal{U} \prec \mathcal{V}$ if and only if for any $V \in \mathcal{V}$ there is a $U \in \mathcal{U}$ such that $V \subseteq U$, so clearly $f^{-1}(V) \subseteq f^{-1}(U)$. Thus $f^{-1}(\mathcal{U}) \prec f^{-1}(\mathcal{V})$ and (4) follows.

Finally, let $\{U_1, U_2, ..., U_{N(\mathcal{U})}\}$ be a subcover of $\mathcal{U}$. Clearly

$$\{f^{-1}(U_1), f^{-1}(U_2), ..., f^{-1}(U_{N(\mathcal{U})})\}$$

is a subcover of $f^{-1}(\mathcal{U})$, so $N(f^{-1}(\mathcal{U})) \leq N(\mathcal{U})$, proving (5). $\square$

We now define the entropy of a continuous map $f : X \to X$ with respect to a cover $\mathcal{U}$.

Definition 1.5. The topological entropy of f relative to $\mathcal{U}$ is given by

$$h(f; \mathcal{U}) = \lim_{n \to \infty} \frac{1}{n} H\left(\bigvee_{i=0}^{n-1} f^{-i}(\mathcal{U})\right) =$$

$$\lim_{n \to \infty} \frac{1}{n} H\Big(\mathcal{U} \vee f^{-1}(\mathcal{U}) \vee f^{-2}(\mathcal{U}) \vee \dots \vee f^{-(n-1)}(\mathcal{U})\Big).$$

Of course, we have to check that this limit exists. Firstly, note that $h(f;\mathcal{U}) \geq 0$ as $H(\mathcal{V}) \geq 0$ for all covers $\mathcal{V}$ of X. Next, we see that the sequence inside the limit is subadditive, i.e. for $m, n \in \mathbb{N}$ we have

$$H\left(\bigvee_{i=0}^{m+n-1} f^{-i}(\mathcal{U})\right) = H\left(\bigvee_{i=0}^{m-1} f^{-i}(\mathcal{U}) \vee \bigvee_{j=m}^{n+m-1} f^{-j}(\mathcal{U})\right)$$

$$\leq H\left(\bigvee_{i=0}^{m-1} f^{-i}(\mathcal{U})\right) + H\left(\bigvee_{j=m}^{n+m-1} f^{-j}(\mathcal{U})\right)$$

by property 2

$$= H\left(\bigvee_{i=0}^{m-1} f^{-i}(\mathcal{U})\right) + H\left(f^{-m}\left(\bigvee_{j=0}^{n-1} f^{-j}(\mathcal{U})\right)\right)$$

by property 3

$$\leq H\left(\bigvee_{i=0}^{m-1} f^{-i}(\mathcal{U})\right) + H\left(\bigvee_{j=0}^{n-1} f^{-j}(\mathcal{U})\right)$$

by property 5.

So $h(f;\mathcal{U})$ exists and is finite for all continuous $f : X \to X$ and open covers $\mathcal{U}$ of X. We may now define topological entropy.

Definition 5.1.5. Let X be a compact topological space, and let $f : X \to X$ be a continuous function. We define the *topological entropy* of f as

$$h(f) = \sup\{h(f;\mathcal{U}) : \mathcal{U} \text{ is an open cover of } X\}.$$

Let X be a compact metric space with metric d, and as before let $f : X \to X$ be a continuous function. The notion of topological entropy using separating and spanning sets can be found in [7] and [17].

Definition 5.1.6. Let $\varepsilon > 0$, and let n be a positive integer. We say that a set $A \subset X$ is (n,ε)-*separated* if for any pair of distinct points $x, y \in A$, we have that there exists an integer k with $0 \leq k \leq n$ such that $d(f^k(x), f^k(y)) \geq \varepsilon$. We define $N_{sep}(n, \varepsilon)$ to be the maximal cardinality of an (n,ε)-separated set in X.

Definition 5.1.7. Let $\varepsilon > 0$, and let n be a positive integer. We say that a set $A \subset X$ (n,ε)-*spans* X if for all $x \in X$ there exists an $y \in A$ such that for all integers k with $0 \leq k \leq n$ we have that $d(f^k(x), f^k(y)) < \varepsilon$. We define $N_{span}(n, \varepsilon)$ to be the minimal cardinality of an (n, ε)-spanning set of X.

Note that the definition of $A \subset X$ being a (n, ε)-spanning set of X is equivalent to the statement that the collection of open balls

$$B(y, \varepsilon) = \left\{ x \in X : \max_{0 \leq k \leq n} d(f^k(y), f^k(x)) < \varepsilon \right\},$$

for every $y \in A$ form an open cover of X. Since X is compact, this means that $N_{span}(n, \varepsilon)$ is finite. We also have the following inequality regarding separating and spanning sets:

Lemma 5.1.8 (Lemma 1 of [7]). *Let $\varepsilon > 0$. Then for any positive integer n, $N_{span}(n, \varepsilon) \leq N_{sep}(n, \varepsilon) \leq N_{span}(n, \frac{\varepsilon}{2})$. Moreover, $N_{sep}(n, \varepsilon)$ is finite.*

Proof. Suppose an (n, ε)-separated set A of maximal cardinality $N_{sep}(n, \varepsilon)$ is not a (n, ε)-spanning set for X. Then by the definition of spanning sets, there exists $x \in X$ such that for all $y \in A$ there is a k with $0 \leq k \leq n$ such that $d(f^k(x), f^k(y)) \geq \varepsilon$, so $A \cup \{x\}$ is also (n, ε)-separated, contradicting the maximality of A. Therefore, A (n, ε)-spans X, so $N_{span}(n, \varepsilon) \leq N_{sep}(n, \varepsilon)$.

Now, let A be an (n, ε)-separated set of maximal cardinality $N_{sep}(n, \varepsilon)$, and let B be an $(n, \frac{\varepsilon}{2})$-spanning set. For the right-hand inequality, we must prove that B has cardinality larger than A. Let $x \in A$, then as B (n, ε)-spans X, there exists a $y \in B$ such that for all k with $0 \leq k \leq n$, $d(f^k(x), f^k(y)) < \frac{\varepsilon}{2}$. Let $g : A \to B$ be defined by $g(x) = y$. It is sufficient to prove that this map is an injection, i.e. that if $g(x_1) = g(x_2)$ then $x_1 = x_2$. Suppose, for a contradiction, there are $x_1, x_2 \in A$ such that $g(x_1) = g(x_2) = c$ but $x_1 \neq x_2$. Then

$$
\begin{aligned}
\epsilon &\leq \max_{0 \leq k \leq n} d(f^k(x_1), f^k(x_2)) \\
&\leq \max_{0 \leq k \leq n} \left[d(f^k(x_1), f^k(c)) + d(f^k(c), f^k(x_2)) \right], \\
&\leq \max_{0 \leq k \leq n} \left[d(f^k(x_1), f^k(c)) \right] + \max_{0 \leq k \leq n} \left[d(f^k(c), f^k(x_2)) \right], \\
&< \frac{\varepsilon}{2} + \frac{\varepsilon}{2}, \\
&< \varepsilon,
\end{aligned}
$$

which is clearly a contradiction. Hence g is injective, so $N_{sep}(n, \varepsilon) \leq N_{span}(n, \frac{\varepsilon}{2})$. This inequality combined with the fact $N_{span}(n, \varepsilon)$ is finite for any $n \geq 1$ and $\varepsilon > 0$ shows that $N_{sep}(n, \varepsilon)$ is also finite. □

Definition 5.1.9. We define the topological entropy of f to be

$$h(f) = \lim_{\varepsilon \to 0} \lim_{n \to \infty} N_{span}(n, \varepsilon)$$
$$= \lim_{\varepsilon \to 0} \lim_{n \to \infty} N_{sep}(n, \varepsilon).$$

Topological entropy is an invariant of topological dynamical systems. Let X be a topological space, and let $f : X \to X$, $g : Y \to Y$ be continuous functions. We say that f and g are *topologically conjugate* if there exists a homeomorphism $\varphi : X \to Y$ such that $\varphi \circ f = g \circ \varphi$. We have the following result:

Theorem 5.1.10. [1] *Let $f : X \to X$ and $g : Y \to Y$ be topologically conjugate mappings. Then*

$$h(f) = h(g).$$

A point p is said to be nonwandering if for every neighbourhood of p there is an integer $n > 0$ such that $f^n(U) \cap V \neq \emptyset$ and the set of all nonwandering points for f is called the *nonwandering set*. If the *nonwandering set* of f is a finite number of periodic points, then $h(f) = 0$.

Some results concerning topological entropy are known for maps of an interval. For instance, for piecewise monotone maps of an interval, the topological entropy is given by $h(f) = \lim_{n \to \infty} \frac{1}{n} \log c_n$, where c_n is the smallest number of intervals on which f^n is monotone [30].

If $f : I \to I$ is an interval map and $s \geq 2$, then an *s-horseshoe* for f is an interval $J \subset I$ and a partition of J into s closed subintervals J_k such that $J \subset f(J_k)$ for $k = 1, \ldots, s$. If f has an horseshoe, then $h(f) \geq \log s$ [2]. We say that f has a constant slope s if on each of its pieces of monotonicity the map is affine with slope s. It is known that for maps of this type $h(f) = max(0, log s)$ [30]. Also if $f : I \to I$ is an interval map, then (see [24] and [25])

$$\limsup_{n \to \infty} \frac{1}{n} \log Card\{x \in I : f^n(x) = x\} \geq h(f).$$

A main question in dynamical systems is the following one raised by John Milnor:

- *Given an explicit dynamical system and given $\epsilon > 0$, is it possible in principle to compute the associated entropy, either topological or measure-theoretic, with a maximum error of ϵ? In practice, is there an effective procedure to carry out this computation in a reasonable length of time?*

When working on topological entropy of one-dimensional Dynamical Systems, one would like to have some simple method of computing it and that is in general a very difficult task. For instance, for piecewise strictly monotone interval maps, various computational methods are known [6] but none of them is really general and simple. Even for Markov maps, before starting computations, which are relatively simple, because one computes only the spectral radius of a nonnegative matrix one has to identify the Markov structure and find the transition matrix. This structure may be complicated and the matrix can be large.

The question of how the topological entropy of a map f depends on the map itself is another major theme in interval dynamics.

Friedman and Tresser [19] showed that for some two-parameter families of circle maps such as the family

$$F_{a,b}(\theta) = \theta + a - \frac{b}{2\pi} \sin 2\pi\theta,$$

the boundary of chaos and the boundary of the parameter region corresponding to a uniquely defined rotation number are not locally connected.

The way they proved their result was by showing that both boundaries of zero topological entropy and zero-width rotation interval have a comb structure for a wide class of two-parameter families of circle maps.

Recall that a comb is a nonlocally connected set, for instance the subset

$$C = B \cap \left(\bigcap_{n \leq 0} T_n \right),$$

of the unit square defined as the union of the base

$$B = \{(x,y)|0 \leq x \leq 1, y = 0\}$$

and the teeth

$$T_n = \{(x,y)|x = 1/n, 0 \leq y \leq 1\},$$

for $n > 0$ and

$$T_0 = \{(x,y)|x = 0, 0 \leq y \leq 1\},$$

Then, for any point $M = (0,y)$ with $y \in (0,1]$ all sufficient small neighbourhoods of M are not connected.

For polynomials of arbitrary degree $d = b+1$ (b-modal polynomials), the following monotonicity statement was proved in [10]. Given $\epsilon \in \{-1, 1\}$ define

$$P_\epsilon^m = \left\{ f : [-1,1] \to [-1,1] : \begin{array}{l} f \text{ is a polynomial with } deg(f) = b+1, \\ f(-1) = \epsilon, f(1) \in \{-1, 1\} \\ f \text{ has b distinct critical points in } (-1, 1). \end{array} \right\}.$$

In the following theorem connectedness of the level sets of topological entropy for polynomial maps is proved (see [10]).

Theorem 5.1.11 (Milnor's monotonicity of entropy conjecture). *For each $h \leq 0$, the isentrope*

$$L_h = \{f \in P_\epsilon^b : h_{top}(f) = h\}$$

is connected.

Moreover, Bruin and van Strien [9] also proved that the isentropes of P_ϵ^b are in a sense rather complicated sets.

Theorem 5.1.12. *Let $f_v \in P_\epsilon^b$ denote the polynomial map with critical values $v = (v_1, v_2, ..., v_b)$. For $b \geq 2$, there are fixed values of $v_2, ..., v_b$ such that the function*

$$v_1 \mapsto h_{top}(f_v),$$

is not monotone.

Theorem 5.1.13. *For any $b \leq 3$ there is a dense set $H \subset [0, \log(b-1)]$ such that for each $b \in H$, the isentrope L_h of P_ϵ^b is not locally connected.*

This last theorem leaves an open conjecture: what happens for $b = 2$ (degree 3).

Buzzi [11] asked the question of whether it can be proved or disproved the upper semicontinuity of entropy for piecewise affine homeomorphisms of the plane.

Let $f : \mathbb{R}^n \to \mathbb{R}^n$ be a homeomorphism, where $n \in \mathbb{Z}^+$. An affine subdivision of f is a finite collection $\mathcal{U} = \{U_1, \ldots, U_n\}$ of pairwise disjoint nonempty open subsets of $\mathbb{R}^n$ such that their union is dense in $\mathbb{R}^n$ and $f|_{U_i} = A_i|_{U_i}$ for each $i = 1, \ldots, N$, where $A_i : \mathbb{R}^n \to \mathbb{R}^n$ is an invertible affine map. A *piecewise affine homeomorphism* is a homeomorphism $f : \mathbb{R}^n \to \mathbb{R}^n$ for which there exists an affine subdivision.

Lozi maps are piecewise affine homeomorphisms of the plane given by

$$\mathcal{L} = \mathcal{L}_{a,b} : \begin{pmatrix} x \\ y \end{pmatrix} \mapsto \begin{pmatrix} 1 - a|x| + by \\ x \end{pmatrix}$$

with $b \in \mathbb{R}$ and $b \neq 0$.

Note that in this case we have $\mathcal{U} = \{U_1, U_2\}$, where $U_1 = \{(x,y) \in \mathbb{R} | x > 0\}$ and $U_2 = \{(x,y) \in \mathbb{R} | x < 0\}$.

Discontinuity of topological entropy for Lozi maps (in fact, a proof that topological entropy for Lozi maps can jump from zero to a value above 0.1203 as one crosses a particular parameter) has been proved by Yildiz [32].

5.2 SCHWARZIAN DERIVATIVE

In spite of the occurrence of topological chaos, what we practically see in an experiment is a periodic motion [26]. In this section, we will study some results regarding the *Schwarzian derivative* of a map. For a more thorough and detailed study of this topic, see [26] and [12].

An important concept in dynamics is the one of attracting periodic points, also sometimes designated *sinks*. An orbit that comes close to an attracting periodic orbit keeps coming closer and closer to it and after some time they cannot be distinguished from it.

In order to show that a periodic point x of period p is attracting, we need to show that $|(f^p)'(x)| < 1$. If, instead, we have that $|(f^p)'(x)| > 1$ then we say that x is repelling.

Consider the interval $J = [0, 1]$ and $J : I \to I$ be a piecewise continuous map (this means that f is continuous and has a finite number of turning points, i.e. a finite number of local extremum). Such a map is called l-modal if f has precisely l turning points and $f(\partial J) \subset \partial J$. We saw that f is unimodal if $f(\partial J) \subset \partial J$ and f has one turning point in the interval.

One of the most important tools in one-dimensional dynamics is the Schwarzian derivative (a really nice condition only using the derivatives of the map) given by:

$$Sf = \frac{f'''}{f'} - \frac{3}{2}\left(\frac{f''}{f'}\right)^2.$$

If we have two maps, f and g such that $Sf < 0$ and $Sg < 0$ then we have that $S(f \circ g) < 0$. Note that $S(f \circ g) = [(Sf) \circ g](g')^2 + Sg$.

An important result regarding Schwarzian derivative is that this condition works for iterates of f, this is, if $Sf < 0$ then $S(f^n) < 0$ for all $n \geq 1$. Note now that

$$S(f^n) = \sum_{i=0}^{n-1} Sf(f^i(x))((f^i)'(x))^2.$$

Note that when we write $Sf(x) < 0$ we mean that negative Schwarzian is zero for x such that $f'(x) \neq 0$. We assume also that the critical points of f are isolated. In [26] we can find a proof for the following theorem:

Theorem 5.2.1. *If $Sf < 0$ then for every periodic attracting orbit there is a critical point of f or an endpoint of $[0, 1]$ which is attracted by this orbit.*

This theorem gives a bound for the possible number of attracting periodic orbits of f with negative Schwarzian derivative.

One of the most remarkable facts in one-dimensional dynamics is the following theorem.

Theorem 5.2.2. *(Singer's Theorem). If f satisfies*

    S1. f is C^1-unimodal,

    S2. f is C^3,

    S3. $Sf(x) < 0$ for all $x \in [-1, 1]$ (at $x = 0$ we allow that $Sf(x) = -\infty$),

then every stable periodic orbit attracts at least one of the points $-1, 0, 1$ (i.e. the end points of the interval or the critical point).

We will follow here the proof in [12].

Let $f, g \in C^3$. As we saw, we have,

$$S(f \circ g)(x) = [(Sf)(g(x))](g'(x))^2 + Sg(x),$$

so the proof of the next lemma is trivial.

Lemma 5.2.3. *If $f \in C^3$ and $Sf(x) < 0$ for all $x \in [-1, 1]$, then $S(f^n)(x) < 0$ for all $x \in [-1, 1]$.*

Lemma 5.2.4. *If $f \in C^3$ and $Sf < 0$ for all x, then $|f'|$ has no positive local minimum in $(-1, 1)$.*

Proof. If f' has an extremum $y \in (-1, 1)$, then $f''(y) = 0$. We have then that

$$Sf(y) = \frac{f'''(y)}{f'(y)},$$

so, $Sf(y) < 0$ implies that $f'''(y)$ and $f'(y)$ have opposite signs.

Thus, f cannot be locally conjugate to, in particular, $x + x^3$, and if $|f'(y)| = 1$ then $|f'| < 1$ on one side of y (locally), this is, every fixed point $y \neq \pm 1$ with $|f'(y)| = 1$ is stable at least from one side. □

Lemma 5.2.5. *Let $f \in C^3$ have finitely many critical points and $Sf(x) < 0$ for all $x \in [-1, 1]$. Then f has only finitely many points of period n for every integer $n \leq 1$.*

Proof. Let $g(x) = f^n(x) = x$ for infinitely many x. Then, by the mean value theorem, $g'(x) = 1$ for infinitely many x.

By Lemmas 5.2.3 and 5.2.4, $|g'|$ has no positive local minimum and therefore it must vanish infinitely often. This contradicts the hypothesis that f, and hence g, has finitely many critical points. □

Lemma 5.2.6. *If $a < b < c$ are consecutive fixed points of $g = f^n$ and if $[a, c]$ contains no critical point of g, then $g'(b) > 1$.*

Proof. By the Mean Value Theorem, there are u, v with $a < u < b < v < c$ such that $g'(u) = g'(v) = 1$. If $g'(x) > 0$ on $[a, c]$, then $g'(b) > 1$ by Lemma 5.2.4.

□

We can now prove Singer's Theorem. Assume $x \in (-1, 1)$ is a stable fixed point for $g = f^n$ and assume that $|g'(x)| < 1$. Define the stable manifold of x as the set of points y for which $g^m(y) \to x$ as $m \to \infty$, and the semilocal stable manifold of x as the connected component of the stable manifold of x, which contains x. If x is in the interior of $[-1, 1]$ this is an open interval (r, s), or it is one of the intervals $[-1, s)$ or $(r, 1]$ (or $[-1, 1]$, a trivial variant).

As in [12] we will consider separately the cases when the semilocal stable manifold is (r, s) or $[-1, s)$.

Let us consider the case (r, s) first. Our function g maps the semilocal stable manifold of x into itself but not r or s into it, since they are not in the stable manifold.

Thus, we have only three possibilities:

(i) $g(r) = r$ and $g(s) = s$,

(ii) $g(r) = s$ and $g(s) = r$,

(iii) $g(r) = g(s) = r$ or $g(r) = g(s) = s$.

Case (i) cannot happen by Lemma 5.2.6, with $r = a, x = b, s = c$ $(r < x < c)$ and case (ii) either, considering g^2 instead of g.

Thus, we have that $g(r) = g(s)$.

By Rolle's Theorem, g has a critical point in (r, s) which is attracted to x.

But if g has a critical point p in (r, s) and $f^n = g$, then a critical point of f is mapped into p.

Let us now consider the case $[-1, s)$. By definition, -1 is attracted to x (we can argue similarly for $(r, 1]$).

This shows Theorem 5.2.2 in all cases when $|g'(x)| < 1$ except if $x = \pm 1$ (in which case there is nothing to prove).

Let us now assume that $g = f^n, g(x) = x$ and $|g'(x)| = 1$. By considering g^2 instead of g, we may assume without loss of generality that $g'(x) = 1$. If $x = \pm 1$ there is nothing to prove.

If $x \in (-1, 1)$, there is, by Lemma 5.2.5, a neighbourhood (r, s) of x containing no other fixed points of g.

Either $g(y) > y$ for all $y \in (r, x)$ or $g(y) < y$ for all $y \in (x, s)$ otherwise $g'(y) > 1$ would have solutions in (r, s) on both sides of x and g' would have a positive local minimum.

So, let us assume that $g(y) > y$ for $y \in (r, x)$. At the minimal value d of y for which $g(y) \geq y$ we have $g(d) = d$ (or $d = -1$ leading to the attraction of the end point). At this point, $g'(d) \geq 1$, and since there is a point w in (d, x) at which $g'(w) = 1$, we get the result in Lemma 5.2.4.

This finishes our proof.

This theorem tells us for instance that if J is a closed interval and $F : J \to J$ a C^3 map with negative Schwarzian derivative and f has n critical points, then f has at most $n + 1$ attracting periodic orbits.

CHAPTER 6

Symbolic dynamics

We define the sequence space Σ_2 on the two symbols 0 and 1 as

$$\Sigma_2 = \{s = (s_0 s_1 s_2 \ldots) | s_j = 0 \text{ or } s_j = 1\}.$$

Consider the space Σ_2, where each point is given by an infinite sequence of 0s and 1s. Take two points on this sequence, say, s and t. We can define the distance between them by the following:

$$d(s,t) = \sum_{i=0}^{\infty} \frac{|s_i - t_i|}{2^i}.$$

This series converges as $|s_i - t_i| = 0$ or $|s_i - t_i| = 1$ and $\sum_{i=0}^{\infty} \frac{1}{2^i} = 2$.

Example 6.0.1. Let $s = (000\ldots)$ and $t = (111\ldots)$, what is $d(s,t)$?

We have

$$d(s,t) = \sum_{i=0}^{\infty} \frac{|s_i - t_i|}{2^i} = \sum_{i=0}^{\infty} \frac{1}{2^i} = 2.$$

Example 6.0.2. Let $s = (000\ldots)$ and t the periodic sequence $t = (010101\ldots)$, what is $d(s,t)$ in this case?

We have

$$d(s,t) = \sum_{i=0}^{\infty} \frac{|s_i - t_i|}{2^i} = \sum_{i=0}^{\infty} \frac{1}{2^{2i}} = \frac{1}{1 - \frac{1}{4}} = \frac{4}{3}.$$

After these two examples, we are now ready to prove the following proposition.

Proposition 6.0.3. *If $s, t \in \Sigma_2$ and $s_i = t_i$ for $i = 0, 1, \ldots, n$, then $d(s,t) \leq \dfrac{1}{2^n}$. Conversely, if $d(s,t) < \dfrac{1}{2^n}$, then $s_i = t_i$ for $i \leq n$.*

Proof. If $s_i = t_i$ for $i \leq n$, then

$$d(s,t) = \sum_{i=0}^{\infty} \frac{|s_i - t_i|}{2^i} = \sum_{i=0}^{n} \frac{|s_i - t_i|}{2^i} + \sum_{i=n+1}^{\infty} \frac{|s_i - t_i|}{2^i} \leq \sum_{i=n+1}^{\infty} \frac{1}{2^i} = \frac{1}{2^n}.$$

If $s_i \neq t_i$ for some $i \leq n$, then

$$d(s,t) = \sum_{i=0}^{\infty} \frac{|s_i - t_i|}{2^i} \geq \sum_{i=0}^{\infty} \frac{1}{2^i} = \frac{1}{2^n}.$$

Thus, if $d(s,t) < \dfrac{1}{2^n}$, then $s_i = t_i$ for $i \leq n$. □

This allows us to decide whether or not two sequences are close to each other (two sequences in our space Σ_2 are close provided their first entries agree).

Before moving to the investigation of our main example, let us introduce the *shift map*. The shift map $\sigma : \Sigma_2 \to \Sigma_2$ is given by

$$\sigma(s_0 s_1 s_2 \ldots) = (s_1 s_2 s_3 \ldots),$$

this is, we delete the first entry and then shift our sequence to the left. Note that the shift map $\sigma : \Sigma_2 \to \Sigma_2$ is continuous.

Now suppose we have a quadratic map where our function is increasing on the left of the critical point (let us denote this interval by I_0) and decreasing on the right (we now denote it by I_1).

In the classical theory of Milnor-Thurston, the symbolic coding is associated to the two intervals where the restriction of the map to each of them is increasing or decreasing (the basic background for symbolic dynamics and kneading theory may be found in [12]).

Definition 6.0.4. The *itinerary* of a point, x under a dynamical system $f(x)$ is given by a sequence $S(x) = s_0 s_1 s_2 \cdots \in \Sigma_2$. Where $s_n = 0$ if $f^n(x) \in I_0$, or $s_n = 1$ if $f^n(x) \in I_1$.

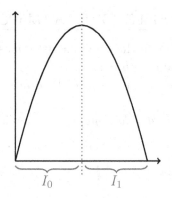 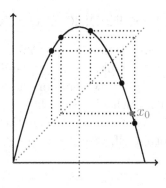

Figure 6.1: [LEFT] The image of a quadratic map $F\mu(x) = \mu x(1 - x)$ over a domain that has been split in half into two intervals, I_0, and I_1. [RIGHT] The same graph with a schematic representation of the first iterations of the point x_0.

For example, for the quadratic map above (see Figure 6.1), we have split the domain between two intervals I_0 and I_1. On the right-hand side, we can see how the point x_0 is mapped under the first few iterations, the itinerary of x_0 here is $S(x_0) = 101011\ldots$, etc.

6.1 KNEADING SEQUENCES FOR DOUBLE STANDARD MAPS

In this section, we give an overview of results on the symbolic dynamics for the double standard family of maps assuming $b = 1$.

We follow some of the work in [5] that was done in collaboration with Michael Benedicks.

As we already saw, in the family of *double standard maps* of the circle onto itself, given by

$$f_{a,b}(x) = 2x + a + (b/\pi)\sin(2\pi x) \pmod 1,$$

where the parameters a, b are real and $0 \le b \le 1$, tongues (sets of parameter values for which there is an attracting periodic point) appear [27, 28].

As we saw, in the classical theory of Milnor-Thurston the symbolic coding is associated to the two intervals where the restriction of the map to each of them is increasing or decreasing (the basic background for symbolic dynamics and kneading theory may be found in [12]). Consider the *double standard family* (1.1) with $b = 1$ and $a \in [0, 1]$, then for each value of the parameter a the map is increasing for all values of $x \in [0, 1]$ except for the values of the parameter for which there is an attractive periodic orbit of period 1 (see [28]).

The explicit computation of the boundary of the period 1 tongue (see [27]) gives:

$$a = \frac{1}{2} \pm \frac{\sqrt{4b^2 - 1} - \arctan\sqrt{4b^2 - 1}}{2\pi}, \tag{6.1}$$

which allows us to compute the interval $[a_0', a_0]$ for which we have an attractive periodic orbit of period one. We get $a_0 \simeq 0.65$, the bifurcation point for the period 1 tongue for $a > 1/2$, and $a_0' = 1 - a_0$.

For $a_0 < a < 1$, f_a has a unique fixed point, which we will denote by $p(a)$. This follows from the bifurcation behaviour of the fixed point(s). For $a = \frac{1}{2}$, f_a has three fixed points, one at $x = \frac{1}{2}$ and two repelling fixed points. According to the Implicit Function Theorem this behaviour persists for $a_0' < a < a_0$. For $a = a_0$, the two rightmost fixed points go through a saddle-node bifurcation and disappear. Since the only bifurcation of fixed points for $1/2 < a < 1$ appears at $a = a_0$ it is clear that there is at most one fixed point in this interval, the continuation of the left fixed point that exists for $\frac{1}{2} < a < a_0$. We denote this fixed point by $p(a)$. We will also prove that $p(a)$ for $a_0 < a < 1$ has a unique preimage different from $p(a)$. We denote this preimage by $q(a)$. For $0 < a < a_0'$ the situation is completely symmetric.

In [5], we used a symbolic coding related to Yoccoz partitions of the interval [33], but in our case we applied it to the circle. Let $J_0 = (p(a), q(a))$ for $a > \frac{1}{2}$, where the circle segments are chosen so that they have positive orientation on the circle, and let $J_1 = \mathrm{int}(\mathbb{T} \setminus J_0)$. In the case $0 < a < a_0'$, $J_0 = [0, q(a)) \cup (p(a), 1)$, where the circle is represented by the half-open fundamental domain $[0, 1)$ and as before J_1 is the interior of its complement.

We may also think of J_0 as the positively oriented (anticlockwise) open arc on the unit circle from $p(a)$ to $q(a)$, and J_1 as being the interior of its complement.

For a given initial point x on the circle such that its orbit does not land on p or q let

$$i_n(x) = \begin{cases} 0 \text{ if } f_a^n(x) \in J_0, \\ 1 \text{ if } f_a^n(x) \in J_1. \end{cases} \tag{6.2}$$

For a point that eventually hits p after possibly passing through q, the coding is so far not defined. For these orbits, we define the coding by either $i_0 \ldots i_n\, 0\,\overline{1}$ or $i_0 \ldots i_n\, 1\,\overline{0}$, and we identify these sequences. Note that this is exactly the same identification as is made of the binary expansions

$$0.i_0 \ldots i_n\, 0\,\overline{1} \text{ and } 0.i_0 \ldots i_n\, 1\,\overline{0},$$

when they are interpreted as real numbers.

Thus, we associate with each $x \in \mathbb{T}$ a finite or infinite sequence of the symbols $0, 1$ called its *itinerary*. We denote by $I(x)$ the sequence $\{i_n(x)\}_{n=0}^{\infty}$ and this sequence is also naturally identified with a real number in $[0, 1]$.

As usual, the kneading sequence will be the itinerary of the critical value $f_a(1/2) = a$ and we denote it by $K(f_a)$. We will sometimes also use the notation $K(a) = K(f_a)$, in particular when we consider the function $a \mapsto K(a)$.

Let $I(x), I(y)$ be two sequences of $\{0, 1\}$ such that $I(x) \neq I(y)$. Furthermore, suppose that

$$I(x) = i_0^1 i_1^1 i_2^1 \cdots i_n^1, \quad I(y) = i_0^2 i_1^2 i_2^2 \cdots i_n^2, \tag{6.3}$$

where i_n is the smallest index for which $i_n^1 \neq i_n^2$.

We order the sequences (6.3) as follows. If $i_n^1 = 0, i_n^2 = 1$ then $I(x) < I(y)$.

This order coincides with the order of the real numbers corresponding to the symbol sequences interpreted as binary expansions.

The map $x \mapsto I(x)$ is continuous, where the topology on the kneading sequences is the topology of real numbers.

In [5] the following lemma was proved.

Lemma 6.1.1. *For $f = f_{a,b}$ with $b = 1$ and for a fixed value of the parameter a, if $x < y$ then $I(x) \leq I(y)$.*

Proof. By the continuity of $x \mapsto I(x)$ it is enough to treat the case when x and y are not binary rational numbers.

We will prove that

$$I(x) > I(y) \implies x > y. \tag{6.4}$$

This result was proved by induction on the index where $I(x)$ and $I(y)$ differ. Clearly, for $n = 0$, (6.4) holds. Assume it holds for $n - 1$ and let us prove it for n. But $I(x) > I(y) \implies I(f(x)) > I(f(y))$. By (6.4), for $n - 1$, we conclude that $f(x) > f(y)$. But by the strict monotonicity of f it now follows that $x > y$. □

As a consequence, we obtain the following result on the monotonicity of the kneading sequences for the family f_a (see [5] for a proof).

Theorem 6.1.2. *Let f_{a_1}, f_{a_2} belong to the double standard family. Then*

(i) $K(f_{a_1}) < K(f_{a_2})$ *implies* $a_1 < a_2$;

(ii) $a_1 < a_2$ *implies* $K(f_{a_1}) \leq K(f_{a_2})$.

Lemma 6.1.3. *Assume that $1/2 < a_1 < 1$, $a_2 = 1 - a_1$ and $K(f_{a_1}), K(f_{a_2})$ are the corresponding kneading sequences interpreted as real numbers. Then $K(f_{a_1}) = 1 - K(f_{a_2})$.*

Proof. Assume that $1/2 < a_1 < 1$ and $a_2 = 1 - a_1$. We start by proving that if $p(a_1), p(a_2)$ are fixed points of f_{a_1}, f_{a_2}, respectively, then $p(a_2) = 1 - p(a_1)$ and $q(a_2) = 1 - q(a_1)$. Consider the lifting F_a of f_a to the real line. A fixed point of f_a satisfies

$$F_a = 2x + a + \frac{1}{\pi}\sin(2\pi x) = x + 1.$$

Consider the map $G(a, x) = F_a(x) - x - 1$. If $(a_1, p(a_1))$ is a solution of $G(a, x) = 0$ then $(1 - a_1, 1 - p(a_1))$ is also a solution. Assume $G_{a_1}(p(a_1)) = p(a_1) + a_1 + \pi^{-1} \sin(2\pi p(a_1)) - 1 = 0$, then $G_{1-a_1}(1 - p(a_1)) = -G_{a_1}(p(a_1)) = 0$.

We prove now that if $p(a_1) < f_{a_1}^n(1/2) < q(a_1)$, then $q(a_2) < f_{a_2}^n(1/2) < p(a_2)$, for $a_2 = 1 - a_1$. We have that

$$p(a_1) < f_{a_1}^n(1/2) < q(a_1),$$

implies

$$1 - p(a_2) < f_{1-a_2}^n(1/2) < 1 - q(a_2),$$

and so

$$q(a_2) < 1 - f_{1-a_2}^n(1/2) < p(a_2).$$

We will prove that

$$f_{a_2}^n(1/2) = 1 - f_{1-a_2}^n(1/2) \tag{6.5}$$

by induction on n. For $n = 1$ we have $f_{a_2}^n(1/2) = 1 - f_{1-a_2}^n(1/2) = a_2 \pmod{1}$. Assume (6.5) holds for n. We show it holds for $n + 1$:

$$f_{a_2}(f_{a_2}^n(1/2)) = f_{a_2}(1 - f_{1-a_2}^n(1/2)) = f_{a_2}(-f_{a_2}^n(1/2))$$
$$= -f_{-a_2}(f_{1-a_2}^n(1/2)) = -f_{1-a_2}(f_{1-a_2}^n(1/2))$$
$$= -f_{1-a_2}^{n+1}(1/2) = 1 - f_{1-a_2}^{n+1}(1/2).$$

Hence we see that for the nth item in the kneading sequences if $k_n(f_{a_1}) = 0$ then $k_n(f_{a_2}) = 1$ and vise-versa so $K(f_{a_1}) = 1 - K(f_{a_2})$. □

Proof of Theorem 6.1.2 for $a < \frac{1}{2}$. We suppose that $0 < a_1 < a_2 < a_0'$. Then $1 - a_2 < 1 - a_1$ and by Theorem 6.1.2 for $a > \frac{1}{2}$ it follows that $K(1 - a_2) \leq K(1 - a_1)$ and (ii) of Lemma 6.1.2 follows. The proof of (i) is similar.

The next lemma will be quite useful. Since $K(a)$ for $0 \leq a \leq a_0'$ and $a_0 \leq a < 1$, may be interpreted as real numbers, it will immediately imply an immediate value theorem for kneading sequences:

Lemma 6.1.4. *The map $a \mapsto K(a)$ is continuous, where the topology on the space of kneading sequences is given by the topology of the real numbers in the interval $[0, 1]$.*

Proof. The argument is standard, see e.g. the proof of [12], Proposition 3.1.2. One sees that $A_0 = \{a|\ K(a) > K_0\}$ and $A_0' = \{a|\ K(a) < K_0\}$ are both open by verifying that each point in these sets is contained in open sets of the type $\{a|\ k_j(a) = k_j^0,\ j = 0, \ldots, n\}$, which in its turn is contained in the corresponding set A_0 and A_0', respectively.

$\square$

6.2 KNEADING SEQUENCES FOR α STANDARD MAPS

In [5] it is studied symbolic dynamics for the double standard family of maps for $b = 1$ and $0 \leq a < 1$. In this section, we will study symbolic dynamics for α-standard maps.

Suppose a map from the α-standard family has an attracting periodic orbit P of period n. The trajectories of the critical points converge to some $p \in P$. Recall from Lemma 2.2.3 that $\varphi_{\alpha,a,b}(p)$ is a periodic point of period n of the α-multiplication map D_α. We will denote this point by $T(P)$ and we call it the *type of the orbit P*. For a periodic point T of D_α we define the *tongue of type T* as the set of parameters (a,b) for which there exists an attracting periodic orbit of type T. Here we will take a modulo 1 and b in the normal range of $[0,1]$. If the period of T is n, we say that the tongue of type T has period n. We want to investigate the order of the tongues as we vary a.

In this section, we are interested in the α-standard maps for $b = 1$ so we will write $f_{\alpha,a,1} = f_{\alpha,a}$ and $\varphi_{\alpha,a,1} = \varphi_{\alpha,a}$.

Lemma 6.2.1. *The interval joining p with $1/2$ is attracted to p under the iterates of $f_{\alpha,a}^n$, where n is the period of P. Thus, if $f_{\alpha,a}$ has an attracting periodic orbit P then $T(P) = \varphi_{\alpha,a}(1/2)$.*

Proof. We have that $f'_{\alpha,a,b}(x) = \alpha + \alpha b \cos(2\pi x)$ and that $1/2$ is a critical point of $f_{\alpha,a,b}$. In particular it is a critical point of $f_{\alpha,a}$.

We now consider the Schwarzian derivative of $f_{\alpha,a}$ which we denote by $(Sf_{\alpha,a})(x)$. After simplifying we have

$$(Sf_{\alpha,a})(x) = \frac{2\pi^2(\cos(2\pi x) - 3)}{1 + \cos(2\pi x)}.$$

We note that the numerator is negative for all x and the denominator is positive for all x except when $x = 1/2$. So for all $x \neq 1/2$ the Schwarzian derivative is negative and $x \to 1/2$, $(Sf_{\alpha,a})(x) \to -\infty$ and so the Schwarzian derivative is negative for all x.

Hence Singer's Theorem [14] applies and as S^1 has no boundary we have that the critical point $1/2$ is in the immediate basin of attraction of p and so we can see the whole interval joining p with $1/2$ is attracted to the periodic orbit P under f_a and so the interval is attracted to p under iterates of $f_{\alpha,a}^n$ where n is the period of P.

To prove the last statement of the lemma we recall the definition of $\varphi_{\alpha,a}(x)$

$$\varphi_{\alpha,a}(x) = \lim_{m \to \infty} \frac{f_a^m(x)}{\alpha^m} \pmod 1$$

which we get from the definition of $\Phi_{\alpha,a}(x)$ and the fact that $\Phi_{\alpha,a}(x)$ is the lift of $\varphi_{\alpha,a}(x)$ to the real line.

We also know that

$$\varphi_{\alpha,a}(f_{\alpha,a}^n(x)) = \alpha^n \varphi_{\alpha,a}(x)$$

as $\varphi_{\alpha,a}$ semiconjugates $f_{\alpha,a}$ with multiplication by α. So we can see that

$$\alpha^n \varphi_{\alpha,a}\left(\frac{1}{2}\right) = \varphi_{\alpha,a}\left(f_{\alpha,a}^n\left(\frac{1}{2}\right)\right) = \lim_{m \to \infty} \frac{f^{nm}\left(\frac{1}{2}\right)}{\alpha^m}.$$

We know that $f_{\alpha,a}^n(\frac{1}{2})$ is attracted to p and that p is periodic of period n thus we have

$$\lim_{m \to \infty} \frac{f_{\alpha,a}^{nm}\left(\frac{1}{2}\right)}{\alpha^m} = \lim_{m \to \infty} \frac{p}{\alpha^m} = \lim_{m \to \infty} \frac{f_{\alpha,a}^{nm}(p)}{\alpha^m} = \varphi_{\alpha,a}\left(f_{\alpha,a}^n(p)\right).$$

Hence using again the fact that $\varphi_{\alpha,a}$ semiconjugates $f_{\alpha,a}$, we have that $\varphi_{\alpha,a}(f_{\alpha,a}^n(p)) = \alpha^n \varphi_{\alpha,a}(p)$. Hence $\alpha^n \varphi_{\alpha,a}(\frac{1}{2}) = \alpha^n \varphi_{\alpha,a}(p)$, which gives that $\varphi_{\alpha,a}(\frac{1}{2}) = \varphi_{\alpha,a}(p)$. We defined $T(P) = \varphi_{\alpha,a}(p)$, which gives the final statement of the lemma, $T(P) = \varphi_{\alpha,a}(\frac{1}{2})$. □

Recall the boundary of the period 1 tongue as in (2.26). Setting $b = 1$ we can compute the interval $[a_0', a_0]$ for which there is an attractive periodic orbit of period 1 (this will be the interval given by the intersection of the period 1 tongue with the $b = 1$ line). In the next lemma, we find two different locations for this interval of a.

Lemma 6.2.2. *The midpoint of the period 1 tongue, which we denote by $[a_0', a_0]$, is at $a = 0$ for α odd and $a = 1/2$ for α even.*

Proof. From (2.26) we can write

$$a = (\alpha - 1)\left(\frac{1}{2} \pm \beta\right),$$

where

$$\beta = \pm \frac{\sqrt{\frac{\alpha^2}{(\alpha-1)^2}b^2 - 1} - \arctan\left(\sqrt{\frac{\alpha^2}{(\alpha-1)^2}b^2 - 1}\right)}{2\pi}.$$

Now let $a_- = (\alpha - 1)(\frac{1}{2} - \beta)$ and $a_+ = (\alpha - 1)(\frac{1}{2} + \beta)$. So we can compute the midpoint $a_{mid} = \frac{a_+ + a_-}{2}$ which gives

$$a_{mid} = \frac{\alpha - 1}{2} = \begin{cases} 0 \pmod 1 & \text{for } \alpha \text{ odd} \\ \frac{1}{2} \pmod 1 & \text{for } \alpha \text{ even.} \end{cases}$$

□

α	a_0'	a_0	a_{mid}
2	0.391	0.609	0.5
3	0.912	0.088	0
4	0.424	0.576	0.5
7	0.943	0.057	0
26	0.471	0.529	0.5

Table 6.1: Values of a_0', a_0 and a_{mid} for various α

Hence, the values of the parameter a for which $f_{\alpha,a}(x)$ has attracting periodic orbits of period 1 depends on α but is always centred in one of two locations. We note also that for α even or odd that $a_0 = a_+$ and $a_0' = 1 - a_0$. Simple substitution into (2.26) gives us Table 6.1.

For maps from the double standard family, for $a_0 < a < 1$, $f_a = f_{a,1}$ has a unique fixed point (as in [5], denoted by $p(a)$). For $a = 1/2$, $f_{a,1}$ had 3 fixed points, one attracting and the other two repelling. In [5] the implicit function theorem is used to show that these 3 fixed points exist for $a_0' < a < a_0$ and that when $a = a_0$ the two rightmost fixed points undergo a saddle-node bifurcation and disappear. So for $a_0 < a < 1$ there is only one fixed point $p(a)$ which is the continuation of the leftmost fixed point that exists for $1/2 < a < a_0$. Also $p(a)$ for $a_0 < a < 1$ has a unique preimage different from $p(a)$ which we will denote by $q(a)$.

Also in [5], the circle is divided according to Yoccoz partitions, this is, $J_0 = (p(a), q(a))$ for $a > 1/2$, and J_1 is the interior of its complement, $J_1 = \text{int}(\mathbb{T} \backslash J_0)$, where $\mathbb{T}$ denotes the circle. In the case that $0 < a < a_0'$, $J_0 = [0, q(a)) \cup (p(a), 1)$ and J_1 defined in the same way. The circle is represented by $[0, 1)$. Defining this partition, a symbolic coding is defined

$$i_n(x) = \begin{cases} 0 \text{ if } f_a^n(x) \in J_0 \\ 1 \text{ if } f_a^n(x) \in J_1 \end{cases}.$$

If the point x eventually hits p perhaps after passing through q we define the coding to be $i_0 \ldots i_n 0\overline{1}$ or $i_0 \ldots i_n 1\overline{0}$. We identify these sequences with each other. We can think of these sequences as binary expansions and the identity above is the same identification as is made of the binary expansions of $0.i_0 \ldots i_n 0\overline{1}$ and $0.i_0 \ldots i_n 1\overline{0}$.

For every x on the circle, we associate it with a finite or infinite sequence of 0s and 1s which we call the *itinerary of x*. We denote the itinerary by $I(x)$, where

$$I(x) = \{i_n(x)\}_{n=0}^{\infty}.$$

The itinerary can be identified with a real number in $[0, 1]$ by taking

$$0.i_0(x)i_1(x)\ldots i_j(x)\ldots.$$

The *kneading sequence* $K(f_a)$ is defined as the itinerary of the critical value of f_a.

By [5], f_a has a stable periodic orbit if and only if $K(f_a)$ is periodic. This is used to show that the periodic part of the binary expansion of $\Phi_a(x)$ gives a periodic coding in 0s and 1s which gives a relation between the periodic kneading sequences and the periodic part of the binary expansion. Namely, if f_a has a periodic orbit of period n for $n > 1$ then $K(f_a)$ corresponds to the shift in the binary expansion of $\Phi_a(1/2) = \frac{k}{2^n-1}$, where $k = 1, \ldots, 2^n - 2$.

This assigns a unique value which earlier we defined as the type of the tongue $T(P)$ to each Arnold tongue and the type of the tongues vary as a increases in the order of the rational numbers $\Phi_a(1/2)$, and so the tongues vary as a increases in the order of the rational numbers corresponding to the kneading sequence $K(f_a)$. For example, for the period 2 tongues we get $\Phi_a(1/2) = 1/3$ and $2/3$, which have corresponding kneading sequence $0.\overline{01}$ and $0.\overline{10}$ respectively.

We would like to be able to replicate this result for the α-standard maps but we run into a problem when trying to split the circle into two intervals J_0 and J_1 as we did above. To rectify this we need to split the interval into more than 2 intervals, in fact into α intervals and we can then impose a base α symbolic coding on $f_a(x)$. So, for example, when $\alpha = 3$ we split the circle into J_0, J_1 and J_2, and the symbolic coding will be as 0s, 1s and 2s.

Lemma 6.2.3. *For α even, if $a \in [0, a_0') \cup (a_0, 1)$ then $f_{\alpha,a}$ has $\alpha - 1$ fixed points and if $a \in [a_0', a_0]$ then $f_{\alpha,a}$ has $\alpha + 1$ fixed points.*

For α odd, if $a \in [0, a_0] \cup [a_0', 1)$ then $f_{\alpha,a}$ has $\alpha + 1$ fixed points and if $a \in (a_0, a_0')$ then $f_{\alpha,a}$ has $\alpha - 1$ fixed points.

Proof. For a fixed point of $f_{\alpha,a}$ we have that $f_{\alpha,a}(x) - x = 0$ and so $(\alpha - 1)x + a + \frac{\alpha}{2\pi}\sin(2\pi x) = 0 \pmod 1$ and so we can write an equation for as

$$a = -(\alpha - 1)x - \frac{\alpha}{2\pi}\sin(2\pi x) \pmod 1. \tag{6.6}$$

We will consider the lift of this to the real line which we will call $G(x)$.

We are interested in the number of fixed points of $f_{\alpha,a}$, thus, if we plot $G(x) \pmod 1$ for $x \in [0, 1)$ and if we draw a horizontal line corresponding to some value of a, the number of times the line crosses the plot of $G(x)$ gives the number of fixed points for that particular value of the parameter a.

This means that when considering $G(x)$ taken modulo 1, every time $G(x)$ wraps around from 0 to 1 it contributes another fixed point to the total number. In terms of the lift, we can see that every time $G(x)$ is equal to an integer we have one more fixed point.

To illustrate this we can think of $G(x)$ as just a straight line given by $-(\alpha - 1)x$. In the case that $\alpha = 2$ we can see that the map ranges only from 0 to -1 and is monotonic and so for $\alpha = 2$, $-(\alpha - 1)x$ has only one fixed point. If $\alpha = 3$ then the map $-(\alpha - 1)x$ would range from 0 to -2 and so when taken modulo 1 would wrap around once and therefore the map would have two fixed points. Now, if we consider $G(x)$ the values which G can take ranges continuously from $G(0) = 0$ to $G(1) = -(\alpha-1)$ and so it is clear for any a such that $0 \le a < 1$ that $f_{\alpha,a}$ has at least $\alpha - 1$ fixed points. Note that it is not exactly $\alpha - 1$ fixed points because $G(x)$ is not monotonic.

In $G(x)$ the sin term also gives fixed points. To show this we notice that $G(x)$ is decreasing for most values of x. We want to know if it is increasing in any region. Solving $G'(x) = 0$ (i.e. solving $(\alpha - 1) + \alpha \cos(2\pi x) = 0$) is precisely the same as solving $F'_{a,b}(x) = 1$. Taking the solution for x modulo 1 we can see that $G(x)$ has stationary points at,

$$x_+ = \frac{1}{2} + \frac{\arctan\left(\sqrt{\frac{\alpha}{(\alpha-1)^2} - 1}\right)}{2\pi},$$

and,

$$x_- = \frac{1}{2} - \frac{\arctan\left(\sqrt{\frac{\alpha}{(\alpha-1)^2} - 1}\right)}{2\pi}.$$

Indeed these are the only stationary points and using the second derivative test we can see that x_- is the local minimum and x_+ the local maximum and so for x such that $x_- < x < x_+$ the map $G(x)$ is increasing. The values of a corresponding to x_- and x_+ are at $a_- = G(x_-)$ and $a_+ = G(x_+)$. Taken modulo 1 we notice that these are precisely the values of a. So $a_+ = a_0$ and $a_- = a'_0$. This means that in this region $[a'_0, a_0]$ the map is increasing and it is the only region where the map is increasing hence, for any particular a in this region, $G(x)$ has 2 more fixed points than occurring over the rest of the map. We have shown that there are at least $\alpha - 1$ fixed points for any particular value of a.

Taking this combined with our knowledge of when the map is increasing or decreasing gives that in the region $[a'_0, a_0]$ there are $\alpha + 1$ fixed points and for any $a \in \mathbb{T} \backslash [a'_0, a_0]$ there are $\alpha - 1$ fixed points.

In Lemma 6.2.2 we wrote the interval as $[a'_0, a_0]$ and so the statement of the lemma for the even case follows from above. We note however in the case that α is odd that a'_0 is greater than a_0 and so to be rigorous we should split the interval and write that in the region $[a'_0, 1) \cup [0, a_0]$ there are $\alpha + 1$ fixed points. $\qquad\square$

Lemma 6.2.3 gives us the tools to begin our study of symbolic dynamics. We have that for $a = a_0$, as a increases, two of the fixed points undergo a saddle-node bifurcation and disappear. For even α there are $\alpha - 1$ fixed points which persist in the $1/2 < a < 1$ region. For α odd these $\alpha - 1$ fixed points persist in the $0 < a < 1/2$ region. We label these as $p_1(a), p_2(a), \ldots, p_{\alpha-1}(a)$ and we let the p_i be ordered such that $p_1 < p_2 < \cdots < p_{\alpha-1}$.

We define the intervals J_i for $i = 0, \ldots, \alpha - 1$ as

$$
\begin{aligned}
J_0 &= [0, p_1), \\
J_1 &= [p_1, p_2), \\
&\ldots \\
J_{\alpha-1} &= [p_{\alpha-1}, 1).
\end{aligned}
\tag{6.7}
$$

This behaviour can be seen in Figures 6.2–6.6. The figures show the values of x along the x axis and a along the y axis for which fixed points of $f_{\alpha,a}$ occur. The dashed lines are at a'_0 and a_0.

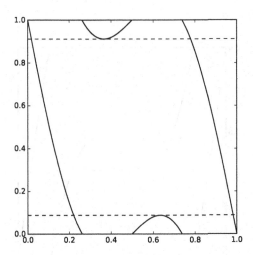

Figure 6.3: The graph of fixed points of $f_{\alpha,a}$ for $\alpha = 3$.

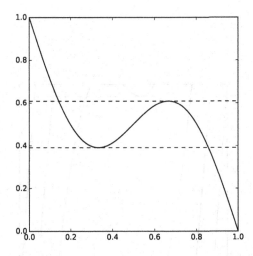

Figure 6.2: The graph of fixed points of $f_{\alpha,a}$ for $\alpha = 2$.

For a given initial point x which does not pass through one of the

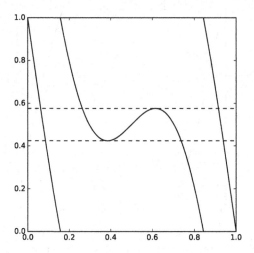

Figure 6.4: The graph of fixed points of $f_{\alpha,a}$ for $\alpha = 4$.

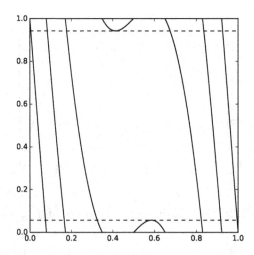

Figure 6.5: The graph of fixed points of $f_{\alpha,a}$ for $\alpha = 7$.

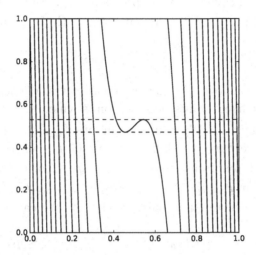

Figure 6.6: The graph of fixed points of $f_{\alpha,a}$ for $\alpha = 26$.

fixed points $p_1, p_2, \ldots p_{\alpha-1}$ we define its *symbolic coding* as follows,

$$
i_n(x) = \begin{cases}
0 \text{ if } f_{\alpha,a}^n(x) \in J_0, \\
1 \text{ if } f_{\alpha,a}^n(x) \in J_1, \\
\cdots \\
i \text{ if } f_{\alpha,a}^n(x) \in J_i, \\
\cdots \\
\alpha - 1 \text{ if } f_{\alpha,a}^n(x) \in J_{\alpha-1},
\end{cases}
$$

and for a point which hits one of the p_is after passing through n other points we define its symbolic coding as $i_0 \ldots i_n (i-1)\overline{\alpha - 1}$ or $i_0 \cdots_n i\overline{0}$ which we identify with each other.

The *itinerary of a point* x is the finite or infinite sequence of 0s, 1s, $\ldots$, $(\alpha-1)$s given by $I(x) = \{I_n(x)\}_{n=0}^{\infty}$. We can then identify a number with the itinerary, given in base α by $0.i_0 \ldots i_n \ldots$.

Theorem 6.2.4. *If a is such that $f_{\alpha,a}$ has a periodic orbit of period n for $n > 1$ then $K(f_{\alpha,a})$ corresponds to the shift in the base α expansion of $\Phi_{\alpha,a}(\frac{1}{2}) = \frac{k}{\alpha^n-1}$, where k is a natural number such that $k = 1, 2, \ldots, \alpha^n - 2$.*

Proof. We set up a linear conjugacy,

$$
T_i(\xi) = \xi + p_i,
$$

where p_i is one of the $\alpha - 1$ fixed points of $f_{\alpha,a}$. We can think of T as a simple rotation by ξ and T conjugates $f_{\alpha,a}$ with $\widehat{f_{\alpha,a}}$, where

$$\widehat{f_{\alpha,a}} = T^{-1} \circ f_{\alpha,a} \circ T,$$

where $\widehat{f_{\alpha,a}}$ is our map $f_{\alpha,a}$ in the new coordinates. We note that in these coordinates our fixed points become points which map to 0, we denote the lift of $\widehat{f_{\alpha,a}}$ by $\widehat{F_{\alpha,a}}$. We want to compute $\Phi_{\alpha,a}(\frac{1}{2}) = \lim_{m \to \infty} \frac{F^m_{\alpha,a}(\frac{1}{2})}{\alpha^m}$. We note that we have $\alpha - 1$ fixed points of $f_{\alpha,a}$ on the circle and so there are $\alpha - 1$ points $q_1 < q_2 < \cdots < q_{\alpha-1}$ such that $\widehat{f_{\alpha,a}}(q_1) = 1$ and in particular $\widehat{F_{\alpha,a}}(q_1) = 1$, $\widehat{F_{\alpha,a}}(q_2) = 2$, $\ldots$, $\widehat{F_{\alpha,a}}(q_{\alpha-1}) = \alpha - 1$ so $\widehat{F_{\alpha,a}}(q_l) = l$, where l is a natural number. We can write $\widehat{F_{\alpha,a}}(x)$ as follows

$$\widehat{F_{\alpha,a}}(x) = \begin{cases} \widehat{f_{\alpha,a}}(x) & 0 \leq x \leq q_1, \\ \widehat{f_{\alpha,a}}(x) + 1 & q_1 \leq x \leq q_2, \\ \ldots & \ldots \\ \widehat{f_{\alpha,a}}(x) + l & q_l \leq x \leq q_{l+1}, \\ \ldots & \ldots \\ \widehat{f_{\alpha,a}}(x) + (\alpha - 1) & q_{\alpha-1} \leq x \leq 1. \end{cases}$$

We recall the fact that for any integer k, $F_{\alpha,a}(x + k) = F_{\alpha,a}(x) + \alpha k$, this still holds for $\widehat{F_{\alpha,a}}(x)$ and so we can write,

$$\widehat{F_{\alpha,a}}(x-k) = \widehat{F_{\alpha,a}}(x) - \alpha k = \begin{cases} \widehat{f_{\alpha,a}}(x - k) & k \leq x \leq q_1 + k, \\ \widehat{f_{\alpha,a}}(x - k) + 1 & q_1 + k \leq x \leq q_2 + k, \\ \ldots & \ldots \\ \widehat{f_{\alpha,a}}(x - k) + l & q_l + k \leq x \leq q_{l+1} + k, \\ \ldots & \ldots \\ \widehat{f_{\alpha,a}}(x - k) + (\alpha - 1) & q_{\alpha-1} + k \leq x \leq k + 1. \end{cases}$$

Adding αk to both sides gives

$$\widehat{F_{\alpha,a}}(x) = \begin{cases} \widehat{f_{\alpha,a}}(x - k) + \alpha k & k \leq x \leq q_1 + k, \\ \widehat{f_{\alpha,a}}(x - k) + \alpha k + 1 & q_1 + k \leq x \leq q_2 + k, \\ \ldots & \ldots \\ \widehat{f_{\alpha,a}}(x - k) + \alpha k + l & q_l + k \leq x \leq q_{l+1} + k, \\ \ldots & \ldots \\ \widehat{f_{\alpha,a}}(x - k) + \alpha k + (\alpha - 1) & q_{\alpha-1} + k \leq x \leq k + 1. \end{cases}$$

Now we want to study what happens to $\widehat{F_{\alpha,a}}^m(x)$ as $m \to \infty$, we set up a recursion formula. Define $x_{j+1} = \widehat{F_{\alpha,a}}(x_j)$ and using our expression for $\widehat{F_{\alpha,a}}(x)$ from above we see that,

$$x_{j+1} = \widehat{F_{\alpha,a}}(x_j) = \widehat{f_{\alpha,a}}(x_j - k) + \alpha k + \begin{cases} 0 & k \le x_j \le q_1 + k, \\ 1 & q_1 + k \le x_j \le q_2 + k, \\ \cdots & \cdots \\ l & q_l + k \le x_j \le q_{l+1} + k, \\ \cdots & \cdots \\ \alpha - 1 & q_{\alpha-1} + k \le x_j \le k + 1. \end{cases}$$

Note that this is our definition of the symbolic coding under our new coordinates and so we can write

$$\widehat{F_{\alpha,a}}(x_j) = \widehat{f_{\alpha,a}}(x_j - k) + \alpha k + i_j.$$

We also notice that under each iteration of our recursion formula the value of k will change so we will say the value of k at iteration j is a_j which gives us that

$$x_{j+1} = \widehat{f_{\alpha,a}}(x_j - a_j) + \alpha a_j + i_j.$$

Note that $\alpha a_j + i_j$ is an integer and when we iterate $\widehat{F_{\alpha,a}}$ again it will become our new k in the equation for $\widehat{F_{\alpha,a}}$ therefore $a_{j+1} = \alpha a_j + i_j$ which gives that $x_{j+1} = \widehat{f_{\alpha,a}}(x_j - a_j) + a_{j+1}$.

We can think of $\widehat{f_{\alpha,a}}(x)$ as a rotation and we will define $\xi_j = \widehat{f_{\alpha,a}}(x_j - a_j)$ and we notice that

$$\xi_{j+1} = \widehat{f_{\alpha,a}}(x_{j+1} - a_{j+1}) = \widehat{f_{\alpha,a}}(\widehat{f_{\alpha,a}}(x_j - a_j) + \alpha a_j + i_j - a_{j+1}) = \widehat{f_{\alpha,a}}(\xi_j).$$

So we get that $x_{j+1} = \xi_{j+1} + a_{j+1}$. We showed earlier that the critical point is attracted to a periodic orbit. If we say the period is n then in particular we can write $x_n = \xi_n + a_n$. This gives us our full recursion formula,

$$\begin{cases} a_{j+1} = \alpha a_j + i_k \\ \xi_{j+1} = \widehat{f_{\alpha,a}}(\xi_j) \end{cases}$$

for $j = 0, 1, 2, \ldots$, where $a_j \in \mathbb{Z}$ and $\xi \in [0, 1)$. This is equivalent to,

$$\begin{cases} x_{j+1} = \widehat{F_{\alpha,a}}(x_j) \\ x_n = a_n + \xi_n. \end{cases}$$

We want to relate this recursion formula to our limit.

We note that $F_{\alpha,a}^m(x)$ when taken modulo 1 is attracted to a periodic orbit so for large m will take one of the values on the orbit of x. In particular, as $m \to \infty$, $F_{\alpha,a}^{mn}(x)$ taken modulo 1 will be attracted to a periodic point determined by $m \pmod{\alpha}$. Every time $m \equiv l \pmod{\alpha}$ for large m, $F_{\alpha,a}^m(x)$ taken modulo 1 will be equal to the same periodic point.

Because of this we consider

$$\lim_{m \to \infty} \frac{F_{\alpha,a}^{mn}(x)}{\alpha^{mn}} = \lim_{m \to \infty} \frac{F_{\alpha,a}^m(x)}{\alpha^m}.$$

We are taking the limit of the critical value for x, our linear conjugacy will move the critical value so that when taking the limit of $\widehat{F_{\alpha,a}}^{mn}(x)$ it will shift then critical value of $f_{\alpha,a}$ to the critical value of $\widehat{f_{\alpha,a}}$ and the so the kneading sequences will be the same for both limits.

Recalling our recursion formula we can say that $\widehat{F_{\alpha,a}}^{mn}(x_0) = x_{mn}$, and so we can write,

$$\Phi_{\alpha,a}(x) = \lim_{m \to \infty} \frac{\widehat{F_{\alpha,a}}^{mn}(x)}{\alpha^{mn}} = \lim_{m \to \infty} \frac{x_{mn}}{\alpha^{mn}} = \lim_{m \to \infty} \frac{a_{mn} + \xi_{mn}}{\alpha^{mn}} = \lim_{m \to \infty} \frac{a_{mn}}{\alpha^{mn}}.$$

To complete the proof it remains for us to find a formula for a_{mn}. We can see that,

$$a_0 = 0, \ a_1 = i_0, \ a_2 = \alpha i_0 + i_1, \ a_3 = \alpha^2 i_0 + \alpha i_1 + i_2, \ldots$$

So we can write $a_n = \alpha^{n-1} i_0 + \alpha^{n-2} i_1 + \cdots + \alpha i_{n-2} + i_{n-1}$ and so,

$$a_{mn} = \alpha^{mn-1} i_0 + \alpha^{mn-2} i_1 + \cdots + \alpha^{(m-1)n} i_{n-1} + \cdots + \alpha i_{kn-2} + i_{kn-1}.$$

Note that because we are considering the iterates of a point which is attracted to a periodic orbit of period n it means that if we iterate by $\widehat{f_{\alpha,a}}^n$ the point will be fixed so it will return to the same region on the circle and so it will have the same symbolic coding. This means that the itinerary is also periodic and will repeat with period n, so $i_n = i_0$, $i_{n+1} = i_1, \ldots$ and so for any m and j we will have $i_{mn+j} = i_j$ so we can write,

$$
\begin{aligned}
a_{mn} = \quad & \alpha^{mn-1} i_0 &+& \quad \alpha^{mn-2} i_1 &+& \ \cdots \ &+& \quad \alpha^{(m-1)n} i_{n-1} &+& \ \cdots \\
+ \quad & \alpha^{(m-1)n-1} i_0 &+& \quad \alpha^{(m-1)n-2} i_1 &+& \ \cdots \ &+& \quad \alpha^{(m-2)n} i_{n-1} &+& \ \cdots \\
+ \quad & \cdots &+& \\
+ \quad & \alpha^{n-1} i_0 &+& \quad \alpha^{n-2} i_1 &+& \ \cdots \ &+& \quad i_{n-1}.
\end{aligned}
$$

Which gives that

$$
\begin{aligned}
a_{mn} = \ &i_0\alpha^{n-1}\left(\alpha^{(m-1)n} + \alpha^{(m-2)n} + \cdots + \alpha^n + 1\right) \\
&+i_1\alpha^{n-2}\left(\alpha^{(m-1)n} + \alpha^{(m-2)n} + \cdots + \alpha^n + 1\right) \\
&+\cdots \\
&+i_{n-1}\left(\alpha^{(m-1)n} + \alpha^{(m-2)n} + \cdots + \alpha^n + 1\right).
\end{aligned}
$$

Since $\sum_{j=0}^{m-1} \alpha^{jn} = \frac{\alpha^{mn}-1}{\alpha^n-1}$, we get

$$
a_{mn} = \frac{\alpha^{mn}-1}{\alpha^n-1}\left(\alpha^{n-1}i_0 + \alpha^{n-2}i_1 + \cdots + i_{n-1}\right).
$$

We substitute this into our equation for the limit $\Phi_a(x)$ and see that,

$$
\begin{aligned}
\Phi_a(x) &= \lim_{m\to\infty} \frac{a_{mn}}{\alpha^{mn}}, \\
&= \lim_{m\to\infty} \frac{\frac{\alpha^{mn}-1}{2^n-1}\left(\alpha^{n-1}i_0 + \alpha^{n-2}i_1 + \cdots + i_{n-1}\right)}{\alpha^{mn}}, \\
&= \lim_{m\to\infty} \frac{1 - \frac{1}{\alpha^{mn}}}{\alpha^n - 1}\left(\alpha^{n-1}i_0 + \alpha^{n-2}i_1 + \cdots + i_{n-1}\right), \\
&= \frac{\alpha^{n-1}i_0 + \alpha^{n-2}i_1 + \cdots + i_{n-1}}{\alpha^n - 1}.
\end{aligned}
$$

We let

$$
k = \alpha^{n-1}i_0 + \alpha^{n-2}i_1 + \cdots + i_{n-1},
$$

different from k in the expression for F_a from above. We want to discover what values k can take. We consider the case where the itinerary of x is not $\overline{0}$ or $\overline{\alpha-1}$ so the smallest value of k we get is when $i_{n-1} = 1$ and all other i_j are equal to 0 which clearly gives $k = 1$. The highest value we can get is when all $i_j = \alpha - 1$ for $j = 0, 1, 2, \ldots, n-2$ and $i_{n-1} = \alpha - 2$. This gives,

$$
\begin{aligned}
k &= \alpha^{n-1}(\alpha-1) + \alpha^{n-2}(\alpha-1) + \cdots + \alpha(\alpha-1) + (\alpha-2), \\
&= \alpha^n - \alpha^{n-1} + \alpha^{n-1} - \alpha^n + \cdots + \alpha^2 - \alpha + \alpha - 2, \\
&= \alpha^n - 2.
\end{aligned}
$$

So we find that k goes from 1 to $\alpha^n - 2$ and in fact k takes every integer value between these. We can see this because our expression for k gives exactly the formula for taking a base α expansion of a number and converting it back to base 10. This gives us the statement of the theorem as if we compute $\frac{k}{\alpha^n-1}$ in base α we get $0.\overline{i_0 i_1 \ldots i_{n-1}}$ which is the decimal number corresponding to our kneading sequence $K(f_a)$.

Recall our definition of the type of a tongue, $T(P) = \varphi_{\alpha,a}(1/2)$, this theorem gives us that as a increases the type of the tongue increases in order of the rational numbers $\frac{k}{\alpha^n-1}$ for $k = 1, 2, \ldots, \alpha^n - 2$. $\qquad \square$

We get the following corollary immediately from Theorem 6.2.4:

Corollary 6.2.5. *At the $b = 1$ level there are $\alpha^n - 2$ Arnold tongues of period n for $n > 1$.*

Example 6.2.6. Using our definition of $\Phi_{\alpha,a,b}$, we now discuss the type of the periods 2 and 3 tongues for a map from the α-standard family. For $\alpha = 3$ as a varies from 0 to 1 the type of the period 2 tongues at the $b = 1$ level is $1/8 < 2/8 < 3/8 < 4/8 < 5/8 < 6/8 < 7/8$ with corresponding kneading sequences $0.\overline{01} < 0.\overline{02} < 0.\overline{10} < 0.\overline{11} < 0.\overline{12} < 0.\overline{20} < 0.\overline{21}$.

As a increases from 0 to 1 the type of the period 3 tongues at the $b = 1$ level is $1/26 < 2/26 < 3/26 < 4/26 < 5/26 < 6/26 < 7/26 < 8/26 < 9/26 < 10/26 < 11/26 < 12/26 < 13/26 < 14/26 < 15/26 < 16/26 < 17/26 < 18/26 < 19/26 < 20/26 < 21/26 < 22/26 < 23/26 < 24/26 < 25/26 <$ with corresponding kneading sequences $0.\overline{001} < 0.\overline{002} < 0.\overline{010} < 0.\overline{011} < 0.\overline{012} < 0.\overline{020} < 0.\overline{021} < 0.\overline{022} < 0.\overline{100} < 0.\overline{101} < 0.\overline{102} < 0.\overline{110} < 0.\overline{111} < 0.\overline{112} < 0.\overline{120} < 0.\overline{121} < 0.\overline{122} < 0.\overline{200} < 0.\overline{201} < 0.\overline{202} < 0.\overline{210} < 0.\overline{211} < 0.\overline{212} < 0.\overline{220} < 0.\overline{221}$.

CHAPTER 7

Tongues

As we saw, one of the main features of the family of standard maps is that the values of the parameters for which there is an attracting periodic orbit are grouped into cusp-like sets called *Arnold tongues*.

For the family of double standard maps, the same is true with some modifications. In this chapter, we will study some features of the tongues in more detail.

Let us introduce some notations. For a fixed b, let us denote the sets of those parameters a for which $f_{a,b}$ has an attracting (resp. neutral) orbit T_b (resp. TN_b). Moreover, let E_b be the set of those parameters a for which $f_{a,b}$ is *expanding*, that is, there exist $C > 0$ and $\varkappa > 0$ such that

$$(f_{a,b}^n)'(x) \geq Ce^{\varkappa n}, \qquad \forall n \geq 0 \quad \forall x \in \mathbb{T}. \tag{7.1}$$

By the result of Mañé [23], if a does not belong to T_b or TN_b, then it belongs to E_b. Observe that by the definition, a small perturbation of an expanding map is also expanding, so E_b is open. In fact, the set $E = \{(a,b) : a \in E_b, 0 \leq b < 1\}$ is open in $[0,1) \times [0,1)$.

Recently, in a joint work with M. Benedicks and M. Misiurewicz, we proved the following result.

Theorem 7.0.1. *For each $b < 1$, the set E_b is dense in the complement of T_b. In particular, every interval of the parameters a either is contained in a closure of one tongue or intersects E_b.*

Proof. Fix $b < 1$. Each tongue is open, so the set T_b is open. Therefore it is the union of countably many components, each of them an open interval. Since the points on the boundary of a tongue belong to TN, and the sets T and TN are disjoint, each component is contained in one tongue.

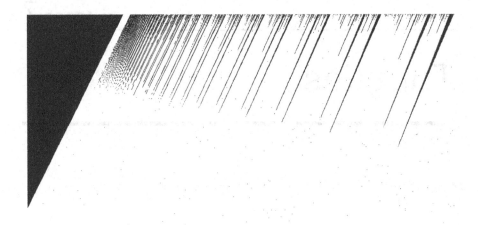

Figure 7.1: Arnold tongues for the family of double standard maps.

We claim that the intersection of the closures of two distinct components A_1 and A_2 is empty. Suppose it is not and that a belongs to this intersection. Then $(a, b) \in TN$, so it has its type. This type must be the same as the type of each of the tongues containing A_1 and A_2, so those types are the same, that is, A_1 and A_2 are contained in the same tongue. If n is the period of the neutral periodic orbit of $f_{a,b}$, the map $f_{a,b}^n$ has an interval on which it looks like one of the Cases 1, 2 or 4 of Lemma 4.1 of [27]. By Theorem 4.1 and Lemma 2.6 of [28], this cannot be Case 4 (a neutral periodic point repelling from both sides), and by Lemma 4.2 of [27] it cannot be Case 1 or 2 (a neutral periodic point repelling from one side). This proves our claim.

If a parameter $a \in TN_b$ does not belong to a boundary of a component of T_b, then by Lemma 4.2 of [27] the neutral periodic orbit of $f_{a,b}$ is repelling from both sides (Case 4), so by Theorem 4.1 and Lemma 2.6 of [28] a is isolated in the set of elements of $T_b \cup TN_b$ which have type of the same period. This proves that there are only countably many such values of a.

By the claim, the complement of T_b is a closed set without isolated points. The set TN_b is countable. Therefore, E_b (which is the complement of T_b minus TN_b) is dense in the complement of T_b.

The second part of the statement follows from the first one and the fact that each component of T_b is contained in one tongue. $\qquad\square$

7.1 LENGTH OF TONGUES

In [27], two particular kinds of orbits were studied in somehow more detail. These were designated by $\mathcal{P}$ and $\mathcal{R}$, respectively, *mostly repelling attracting periodic orbits* and *intermittent periodic orbits*.

Before describing these two types of orbits let us recall some results we already investigated for a-standard maps.

Assume that $F_{a,b}$ are maps from the real line to itself, satisfying the following properties:

1. Each $F_{a,b}$ is continuous increasing (as a function of x),

2. $F_{a,b}(x+k) = F_{a,b}(x) + 2k$ for every integer k,

3. $F_{a,b}(x)$ is increasing as a function of a and continuous jointly in x, a, b.

While the fact that local homeomorphisms of the circle of degree 2 are semiconjugate to the doubling map is well known, we need additionally monotonicity properties of the semiconjugacy as the function of a. Therefore, we include a simple proof which also gives us this monotonicity.

The next two lemmas establishes semiconjugacy as a certain limit. See Section 3 in [27] for a detailed proof.

Lemma 7.1.1. *Under the assumptions (1) and (2), the limit*

$$\Phi_{a,b}(x) = \lim_{n \to \infty} \frac{F_{a,b}^n(x)}{2^n} \tag{7.2}$$

exists uniformly in x. The limit $\Phi_{a,b}(x)$ is a continuous increasing function of x. Moreover, $\Phi_{a,b}(x+k) = \Phi_{a,b}(x) + k$ for every integer k and $\Phi_{a,b}(F_{a,b}(x)) = 2\Phi_{a,b}(x)$ for every x, so $\Phi_{a,b}$ semiconjugates $F_{a,b}$ with multiplication by 2.

Lemma 7.1.2. *Under the assumptions of Lemma 7.1.1, the map $\Phi_{a,b}$ is a lifting of a monotone degree one map $\varphi_{a,b}$ of the circle to itself, which semiconjugates $f_{a,b}$ with the doubling map $D : x \mapsto 2x \pmod 1$. Moreover, if p is a periodic point of $f_{a,b}$ of period n then $\varphi_{a,b}(p)$ is a periodic point of D of period n.*

The class $\mathcal{P}$ of attracting periodic orbits are attracting periodic orbits for f_a of type

$$0.0001 * 1 * 1 \cdots * 1$$

(the line over a finite sequence means that it is repeated periodically), where each $*$ can be 0 or 1.

For the values of a,

$$a_l \approx -0.32221099$$

and

$$a_r \approx -0.28609229$$

we have, respectively, $\Phi_{a_l}(1/2) = 1/16$ and $\Phi_{a_r}(1/2) = 1/8$.

We have $1/16 = 0.0001\overline{0}$ and $1/8 = 0.000\overline{1}$. The numbers of the form $0.0001 * 1 * 1 \cdots * 1$ are between those two, so any a for which f_a has a periodic orbit of such type is in (a_l, a_r).

Let us consider now periodic orbits of the type $\mathcal{R}$. Let $b = 1$ and $f_a = f_{a,1}$ with lifting $F_a = F_{a,1}$. If

$$a_I = \frac{\sqrt{3}}{2\pi} - \frac{2}{3} \sim -0.3910022190,$$

then

$$F_{a_I}\left(\frac{2}{3}\right) = \frac{2}{3}$$

and

$$F'_{a_I}\left(\frac{2}{3}\right) = 1.$$

Thus, $2/3$ is a neutral fixed point and if a is a little larger than a_I, then we have *intermittency* for f_a. The trajectories of points in a large interval containing $1/2$ are increasing and spend a lot of time very close to $2/3$.

As in [27] we denote by $\mathcal{R}$ the class of attracting periodic orbits for f_a such that if $p \in P \in \mathcal{R}$ then $1/2$ is in the immediate basin of attraction of p and n is the period of P then

$$p < F_a(p) < F_a^2(p) < F_a^3(p) < \cdots < F_a^{n-1}(p),$$

with $p = F_a^n(p) - 1$.

Figure 7.1 shows the tongues of period 50 or less in the intermittent region.

As in [27] we define *proper tongues* as the components of the tongues which intersect the line $b = 1$. By results in [27], the intersection of any tongue with the line $b = 1$ is connected and nonempty. Thus, there is exactly one proper tongue of each type.

We measure the length of a tongue in the direction of b. The first

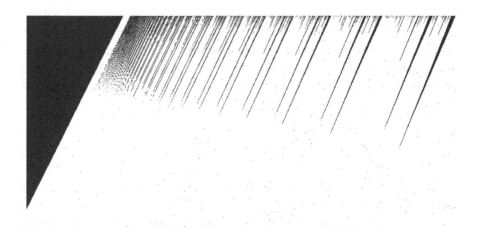

Figure 7.2: Tongues of period 50 or less in the intermittent region, $0.6 \leq a \leq 0.64$, $0.96 \leq b \leq 1$.

result seems to confirm the conjecture that at a given level $b < 1$ there are only finitely many tongues.

The next two theorems were proved in [27].

Theorem 7.1.3. *Let s, t be periodic points of D with $1/16 < s < t < 1/8$. Then there exist constants $\lambda > 1$, $N > 0$ and $K_5 > 0$ such that any proper tongue of a type between s and t, period $n \geq N$, and such that the orbit of this type for some f_a belongs to $\mathcal{P}$, has a length smaller than $K_5 \lambda^{-n}$.*

Let us now consider periodic orbits of the type $\mathcal{R}$. Here we will see that if $b < 1$ is sufficiently close to 1 then there are infinitely many tongues at that level. Once we know where they are situated, we can produce a picture showing them (see Figure 7.2). Let us remark that a straightforward method used to detect attracting periodic orbits does not work well here, since a point that moves only slightly due to intermittency may be mistaken for a fixed point.

Theorem 7.1.4. *There exists a constant $L > 0$ such that any proper tongue such that the orbit of this type for some f_a belongs to $\mathcal{R}$, has length larger than L.*

7.2 BOUNDARY OF THE TONGUES

Throughout this book, we have been working with real analytic increasing maps of the circle. In this section, we investigate the shape of the boundaries of the tongues, in particular close to their tips. We follow closely the work done in [28].

Recall that we call an orbit

$$(x, f(x), \ldots, f^{n-1}(x)),$$

with $f^n(x) = x$, *attracting* if $(f^n)'(x) < 1$. Moreover, if $(f^n)'(x) > 1$ we call it *repelling* and *neutral* if $(f^n)'(x) = 1$.

In a sense, this is what we designate by *differentiably* attracting, repelling and neutral, respectively.

If there is $\varepsilon > 0$ such that $f^n(y) < y$ for $y \in (x, x + \varepsilon)$, then we call x *topologically attracting from the right*.

In the same way, we define periodic orbits *topologically attracting from the left*, and *topologically repelling from the right (left)*.

Moreover, an orbit topologically attracting (repelling) from both sides is just *topologically attracting (repelling)*. Since maps from our double standard family are analytic and not equal to the identity (we can assume this because they have degree 2), each periodic orbit is either topologically attracting or topologically repelling from each side.

One of the main tools used to study tongues in the family of double standard maps is Theorem 3.0.7 (see also Theorem 3.5 of [27]).

We have the following result from [27].

Lemma 7.2.1. *A neutral periodic orbit mentioned in Theorem 3.0.7 cannot be topologically attracting from both sides.*

Proof. If a neutral periodic orbit is topologically attracting from both sides, it has at least two immediate basins of attraction that are not symmetric with respect to each other (but each of them is symmetric itself). Thus, such a situation is also impossible. □

Another important tool from [27] is the monotone degree one circle map $\varphi_{a,b}$, which semiconjugates $f_{a,b}$ with the doubling map $D : x \mapsto 2x$ (mod 1). It is used in one of the key lemmas about the family of double standard maps (this is Lemma 4.1 of [27]).

Lemma 7.2.2. *Assume that p is an attracting or neutral periodic point of $f_{a,b}$ of period n. Let J be the set of all points x for which $\varphi_{a,b}(x) =$*

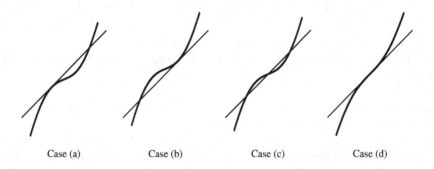

Case (a) Case (b) Case (c) Case (d)

Figure 7.3: Four cases from Lemma 7.2.2.

$\varphi_{a,b}(p)$. *Then J is either a closed interval (modulo 1) or a singleton and $f_{a,b}^n|_J$ is an orientation preserving homeomorphism of J onto itself. The endpoints of J are fixed points of $f_{a,b}^n$, and one of the following four possibilities holds. In the first three cases J is an interval.*

1. *The left endpoint of J is neutral, topologically attracting from the right and topologically repelling from the left; the right endpoint of J is repelling; there are no other fixed points of $f_{a,b}^n$ in J.*

2. *The right endpoint of J is neutral, topologically attracting from the left and topologically repelling from the right; the left endpoint of J is repelling; there are no other fixed points of $f_{a,b}^n$ in J.*

3. *Both endpoints of J are repelling; there is an attracting fixed point of $f_{a,b}^n$ in the interior of J; there are no other fixed points of $f_{a,b}^n$ in J.*

4. *The set J consists of one neutral fixed point of $f_{a,b}^n$, repelling from both sides.*

Let us look at the four possibilities carefully, distinguishing between differentiable and topological properties. In Cases 1 and 2 one endpoint of J is neutral, so the other one has to be (differentiably) repelling. In Case 3 the fixed point in the interior of J is topologically attracting, so by Lemma 7.2.1 it is (differentiably) attracting. Then the endpoints have to be (differentiably) repelling. In Case 4 the fixed point is neutral, so "repelling" means here "topologically repelling."

Our parameter space is $\mathbb{T} \times [0, 1]$, and when we speak of the *boundary* of a tongue, we mean the boundary relative to this space. In particular, if a tongue contains a segment of the line $b = 1$, only the endpoints of this segment belong to the boundary of the tongue.

Regarding the boundary of the tongues, we have the following result from [28].

Lemma 7.2.3. *If (a_0, b_0) belongs to the boundary of a tongue of period n then $f^n_{a_0,b_0}$ has a neutral fixed point.*

Proof. If (a_0, b_0) belongs to the boundary of a tongue of period n then there is a sequence of points (a_n, b_n) convergent to (a_0, b_0), such that f_{a_n,b_n} has an attracting periodic point x_n for all $n \geq 1$. By compactness of $\mathbb{T}$, we may assume that x_n converges to some x_0 as $n \to \infty$. Since $f_{a,b}(x)$ and $f'_{a,b}(x)$ depend continuously on (a, b, x), we see that x_0 is an attracting or neutral fixed point of $f^n_{a_0,b_0}$. Tongues are open, and by Theorem 3.0.7 they are pairwise disjoint. Thus, a point on the boundary of a tongue cannot belong to any tongue. Therefore, x_0 cannot be attracting for $f^n_{a_0,b_0}$, so it has to be neutral. □

Also from [28] we have the following definition and the next lemmas.

Definition 7.2.4. We will say that a point (a, b) from a boundary of a tongue of period n belongs to the *left boundary of the tongue* if there is $\varepsilon > 0$ such that $(a + t, b)$ belongs to the tongue for $0 < t < \varepsilon$ and does not belong to the tongue for $-\varepsilon < t < 0$. Similarly, a point (a, b) from a boundary of a tongue of period n belongs to the *right boundary of the tongue* if there is $\varepsilon > 0$ such that $(a + t, b)$ belongs to the tongue for $-\varepsilon < t < 0$ and does not belong to the tongue for $0 < t < \varepsilon$. Moreover, a point (a, b) from a boundary of a tongue of period n is the *tip* of this tongue if there is $\varepsilon > 0$ such that $(a + t, b)$ does not belong to the tongue for $-\varepsilon < t < \varepsilon$.

In the classification of Lemma 7.2.2 if 3 is satisfied then the point (a, b) belongs to a tongue of period n. If (a, b) belongs to a boundary of a tongue of period n then, by Lemma 7.2.3, $f^n_{a,b}$ has a neutral fixed point. Thus, one of the Cases 1, 2 or 4 holds. We will investigate how those cases are related to the notions in Definition 7.2.4. We start with Cases 1 and 2.

Let us first show that the partial derivative of $f^n_{a,b}$ with respect to a is positive.

Lemma 7.2.5. *For any $n \geq 1$, $x \in \mathbb{T}$, we have*

$$\frac{\partial f_{a,b}^n}{\partial a}(x) > 0. \tag{7.3}$$

Proof. Since

$$f_{a,b}^n(x) = f_{a,b}^{n-1}(f_{a,b}(x)),$$

we have

$$\frac{\partial f_{a,b}^n}{\partial a}(x) = \frac{\partial f_{a,b}^{n-1}}{\partial a}(f_{a,b}(x)) + (f_{a,b}^{n-1})'(f_{a,b}(x))\frac{\partial f_{a,b}}{partiala}(x).$$

since, $\frac{\partial f_{a,b}}{\partial a}(y) = 1$ for every y, by induction we obtain

$$\frac{\partial f_{a,b}^n}{\partial a}(x) = \sum_{k=0}^{n-1}(f_{a,b}^k)'(f_{a,b}^{n-k}(x)).$$

Since $f_{a,b}'$ is nonnegative everywhere, so is $(f_{a,b}^k)'$. Moreover, if $k = 0$ then $(f_{a,b}^k)' \equiv 1$. This proves (7.3). □

Lemma 7.2.6. *In Case 1 of Lemma 7.2.2 (a,b) belongs to the left boundary of a tongue of period n. Similarly, in Case 2 (a,b) belongs to the right boundary of a tongue of period n.*

Proof. Assume that Case 1 holds. Let x be a neutral fixed point of $f_{a,b}^n$. Then $f_{a,b}^n(x) = x$. Since x is attracting from the right and $f_{a,b}^n$ is increasing, we have $f_{a,b}^n(y) < y$ for some $y > x$ (here we do not take the map $f_{a,b}^n$ modulo 1). Thus, by (7.3), there exists $\varepsilon > 0$ such that if $0 < t < \varepsilon$ then $f_{a+t,b}^n(x) > x$ and $f_{a+t,b}^n(y) < y$. Therefore, between x and y the map $f_{a+t,b}^n$ has a fixed point which is topologically attracting from both sides. It cannot be neutral by Lemma 7.2.1. Thus, it is attracting. This proves that $(a + t, b)$ belongs to a tongue of period n.

Since x is attracting from the right and repelling from the left, and $f_{a,b}^n$ is increasing, there exists $\delta > 0$ such that $f_{a,b}^n(z) \leq z$ for all $z \in (x - \delta, x + \delta)$. Again by (7.3), there exists $\varepsilon_1 > 0$ such that if $0 < t < \varepsilon_1$ then $f_{a-t,b}^n$ has no fixed points in $(x - \delta, x + \delta)$. The same argument can be used for the remaining $n - 1$ points from the $f_{a,b}$-orbit P of x. In such a way we find $\varepsilon_2 > 0$ and a neighborhood U of P such that if $0 < t < \varepsilon_2$ then $f_{a-t,b}^n$ has no fixed points in U. In $\mathbb{T} \setminus U$ the map $f_{a,b}^n$ has only finite number of fixed points, all of them repelling. Therefore, if ε_2 is sufficiently small, then $f_{a-t,b}^n$ has no attracting fixed points in $\mathbb{T} \setminus U$.

This proves that there is $\varepsilon_3 > 0$ such that if $0 < t < \varepsilon_3$ then $(a - t, b)$ does not belong to a tongue of period n. Hence, (a, b) belongs to the left boundary of a tongue of period n.

The proof that in Case 2 (a, b) belongs to the right boundary of a tongue of period n is similar. □

7.3 TIP OF THE TONGUES

Consider now a family of maps of an interval or a circle that looks locally more or less like the family of double standard maps, also assume that the maps are real analytic and that they depend on the parameters in a real analytic way. We will show that if the dynamics of a map in this family locally look like the one observed at the tip of a tongue and the parameters belong to the closure of a tongue, then it is really the tip. We will consider here only one parameter, whose change moves the graph of the map up or down. Note that the analyticity of the function plays an essential role in the proof.

Lemma 7.3.1. *Let U be a neighbourhood of the origin in $\mathbb{R}^2$ and let $G : U \to \mathbb{R}$ be a real analytic function. Assume that*

$$G(0, x) < 0 \ \text{for} \ x < 0, \quad G(0, 0) = 0, \quad \text{and} \ G(0, x) > 0 \ \text{for} \ x > 0; \tag{7.4}$$

and that

$$\frac{\partial G}{\partial t}(0, 0) \neq 0, \tag{7.5}$$

where t is the first variable. Then there are open intervals I, J containing 0 such that $I \times J \subset U$ and for every $t \in I$ is exactly one $x \in J$ such that $G(t, x) = 0$. Moreover, for those t and x, if $t \neq 0$ then $\frac{\partial G}{\partial x}(t, x) > 0$.

Proof. By (7.5) we can apply the real analytic Implicit Function Theorem. Hence, there exists a real analytic function ψ, defined in an open interval I containing 0, such that $G(\psi(x), x) = 0$ for all $x \in I$. Moreover, these are the only solutions of the equation $G(t, x) = 0$ in a small neighbourhood of the origin. If ψ is constant, then, since $G(0, 0) = 0$, it is 0, so $G(0, x) = 0$ for x close to 0. This contradicts (7.4). Thus, ψ is not constant. Since ψ is analytic, this implies that if I is sufficiently small then

1. $\psi'(x) \neq 0$ for all $x \in I \setminus \{0\}$.

Set $J = \psi(I)$. We may assume that I is so small that furthermore

2. $J \times I \subset U$;

3. in $J \times I$ all solutions to $G(t, x) = 0$ are given by $t = \psi(x)$;

4. $\frac{\partial G}{\partial t}$ has a constant sign in $J \times I$.

Assume that the sign mentioned in 4 is $+$. Then, by (7.4), if $G(t, x) = 0$ then x and t have opposite signs. By 1, in each component of $I \setminus \{0\}$ the function ψ is strictly monotone. Those two facts, together with $\psi(0) = 0$, imply that ψ is strictly decreasing in I (in particular, $\psi'(x) < 0$ for all $x \in I \setminus \{0\}$). Therefore, taking 3 into account, we see that if $t \in I$ then there is exactly one $x \in J$ (namely, $x = \psi^{-1}(t)$) such that $G(t, x) = 0$.

We have

$$0 = \frac{d}{dx} G(\psi(x), x) = \frac{\partial G}{\partial t}(\psi(x), x) \cdot \psi'(x) + \frac{\partial G}{\partial x}(\psi(x), x),$$

so since $\frac{\partial G}{\partial t} > 0$ and $\psi'(x) < 0$, we get $\frac{\partial G}{\partial x}(\psi(x), x) > 0$ for all $x \in I \setminus \{0\}$.

If the sign mentioned in 4 is $-$, the proof is the same, except that if $G(t, x) = 0$ then x and t have the same signs, so ψ is strictly increasing and $\psi'(x) > 0$ for $x \in I \setminus \{0\}$, and thus we reach the same conclusion. $\square$

As a corollary we get the main result of this section.

Theorem 7.3.2. *Let U be a neighborhood of the origin in $\mathbb{R}^2$ and let $F : U \to \mathbb{R}$ be a real analytic function. Set $f_t(x) = F(t, x)$. Assume that f_0 has a topologically repelling fixed point at $x = 0$ and that*

$$\frac{\partial F}{\partial t}(0, 0) \neq 0. \tag{7.6}$$

Then there are open intervals I, J containing 0 such that $I \times J \subset U$ and for every $t \in I$ the map f_t has exactly one fixed point $x \in J$. Moreover, if $t \neq 0$ then this fixed point is (differentiably) repelling.

Proof. Set $G(t, x) = F(t, x) - x$ and apply Lemma 7.3.1. $\square$

The intuitive interpretation of the conclusion of this theorem is that the point $(0, 0)$ in the parameter space can be neither in the interior of a tongue nor in its left or right boundary. Thus, if it is in the closure of a tongue, it has to be its tip.

7.3.1 Applications of Theorem 7.3.2 to double standard maps

As we apply Theorem 7.3.2 to the family of double standard maps, we get the following result.

Theorem 7.3.3. *Consider a one-parameter subfamily $g_t = f_{a(t),b(t)}$ of the family of double standard maps, where $a(t)$ and $b(t)$ depend on t in an analytic way. Assume that $g_{t_0}^n$ has a topologically repelling neutral fixed point x_0 and $\partial G/\partial t(t_0, x_0) \neq 0$, where $G(t, x) = g_t^n(x)$. Then there is $\varepsilon > 0$ such that if $|t - t_0| < \varepsilon$ then g_t^n has no attracting or neutral fixed point.*

Proof. By Theorem 7.3.2 there is $\varepsilon_1 > 0$ and a neighborhood U_1 of x_0 such that if $|t - t_0| < \varepsilon_1$ then g_t^n has no attracting or neutral fixed point in U_1. As in the proof of Lemma 7.2.6 we use the same argument for the remaining $n - 1$ points of the g_{t_0}-orbit P of x_0 to get $\varepsilon_2 > 0$ and a neighborhood U of P such that if $|t - t_0| < \varepsilon_2$ then g_t^n has no attracting or neutral fixed point. In $\mathbb{T} \setminus U$ the map $g_{t_0}^n$ has only finite number of fixed points, all of them repelling, so the same holds for g_t^n when t is sufficiently close to t_0. This completes the proof. □

By Theorem 7.3.3 and Lemma 7.2.6 we get immediately the following theorem for the family of double standard maps.

Theorem 7.3.4. *In Case 1 of Lemma 7.2.2, the point (a, b) belongs to the left boundary of a tongue of period n, in Case 2 to the right boundary of a tongue of period n, and in Case 4 it is a tip of a tongue of period n.*

This theorem has important consequences. Basically, it yields that the tongues have regular, tongue-like shapes. The first corollary is immediate.

Corollary 7.3.5. *Every point on the boundary of a tongue either belongs to its left or right boundary or is its tip.*

In particular, degeneracies like a horizontal segment contained in the boundary of a tongue are ruled out.

Remember, that the definition of a tongue we adopted in this paper automatically makes tongues connected.

Corollary 7.3.6. *The intersection of every tongue with any horizontal line $b = $ constant is connected. In particular, every tongue is simply connected.*

Proof. If the intersection of a tongue of period n with a horizontal line is not connected, then, since the tongue itself is connected, there is c such that the intersection of the tongue with the horizontal line $b = c$ contains a point (a, c) belonging to the closure of two different components of this intersection. In this situation, small changes of a in both directions produce an attracting periodic orbit of period n. Therefore, (a, c) is not a tip and does not belong to either left or right boundary of the tongue, a contradiction. □

The next corollary follows immediately from Theorem 7.3.4 and Proposition 4.6 of [28] (which states that whenever a piece of the boundary of a tongue consists of points for which the Case 1 or 2 of Lemma 7.2.2 holds, it has slope with the absolute value at least π).

Corollary 7.3.7. *The left and right boundaries of a tongue have slope with an absolute value of at least π.*

We can get even more information about the shape of a tongue at its tip.

Theorem 7.3.8. *At a tip of a tongue, the left and right boundaries are tangent to each other.*

Proof. Let (a_0, b_0) be a tip of a tongue of period n, and let x_0 be the corresponding neutral periodic point of period n of f_{a_0,b_0}. By Theorem 7.3.4, Case 4 of Lemma 7.2.2 applies to it. Fix this x_0 and consider the function $\psi(a, b) = f_{a,b}^n(x_0)$. By Lemma 7.2.5, $\frac{\partial \psi}{\partial a} > 0$, and therefore, the gradient vector v of ψ is nonzero. The only vectors in direction of which the derivative of ψ is 0 are orthogonal to v. When moving parameters a, b in the direction of any other vector (α, β), starting from (a_0, b_0), we get immediately outside of the tongue by Theorem 7.3.3 applied to $g_t = f_{a+t\alpha, b+t\beta}$. Thus, v must be normal to both left and right boundaries of the tongue at (a_0, b_0). Therefore, those boundaries are tangent to each other. □

7.4 CONNECTEDNESS OF TONGUES

One important question about the tongues for double standard maps is whether they are connected. This question was solved in [16].

Let us start with the following lemma which tells us that each $f_{a,b}$ is monotonic semiconjugate to the doubling map $D : x \mapsto 2x \bmod 1$ via a unique mapping $\varphi_{a,b}$.

This lemma also appears in [16] and we will follow the proof there.

Lemma 7.4.1. *Let $f : \mathbb{T}^1 \to \mathbb{T}^1$ be a monotonic continuous mapping. If $\varphi, \psi : \mathbb{T}^1 \to \mathbb{T}^1$ are nondecreasing continuous mappings of degree 1 such that $\varphi \circ f = 2 \times \varphi$ and $\psi \circ f = 2 \times \psi$, then $\varphi = \psi$.*

Proof. Let F be a lift of f and $\tilde{\varphi}$ a lift of φ. Since $\tilde{\varphi} \circ F$ is a nondecreasing real function, $\tilde{\varphi} \circ F$ is a lift of 2φ.

Thus, there exists an integer k such that

$$\tilde{\varphi}(F(x)) = 2\tilde{\varphi}(x) + k$$

for all $x \in \mathbb{R}$.

It follows that if we take $\varphi_1 = \tilde{\varphi} + k$, then we have $\varphi_1 \circ F = 2\varphi_1$.

Assume here we have the same properties for a lift ψ_1 of ψ.

The mappings φ_1 and ψ_1 are both nondecreasing of degree 1 and locally bounded, thus, the function $\varphi_1 - \psi_1$ is periodic and bounded. But then

$$\varphi_1(x) - \psi_1(x) = \frac{1}{2} \left(\varphi_1(F(x)) - \psi_1(F(x)) \right),$$

so

$$\varphi_1(x) - \psi_1(x) = \left(\frac{1}{2}\right)^n \left(\varphi_1(F^{\circ n}(x)) - \psi_1(F^{\circ n}(x)) \right)$$

for all n. Consequently,

$$\varphi_1(x) - \psi_1(x) = 0.$$

$\square$

Using quasiconformal techniques, A. Dezotti proved the following beautiful result:

Theorem 7.4.2. *For each type τ (periodic point of D) there exists a unique $(a_\tau, b_\tau) \in T_\tau$ such that f_{a_τ, b_τ} has a superattracting cycle. Then $b_\tau = 1$ and all $(a, b) \in T_\tau$ can be joined to $(a_\tau, 1)$ by a continuous path in T_τ.*

and from this we have the following corollary:

Corollary 7.4.3. *The tongues are connected.*

In the above theorem Dezotti defines $(a, b) \in \mathbb{T}^1 \times [0, 1]$ as being of type τ if $g_{a,b}$ has an attracting cycle on the circle with $\varphi_{a,b}(x_0) = \tau$, where x_0 is the distinguished point of the attracting cycle of $g_{a,b}$ and a tongue T_τ being of type $\tau \in \mathbb{T}^1$ as the set of parameters $(a, b) \in \mathbb{T}^1 \times [0, 1]$ of type τ.

7.5 ARNOLD TONGUES OF HIGHER PERIODS FOR α-STANDARD MAPS

In this section, we want to study both the Arnold tongues of period 1 and higher. In particular, we want to look at the limit $\varphi_{a,b}$ and how this relates to the tongues. We build up some machinery to study this relationship.

Suppose a map from the α-standard family has an attracting periodic orbit P of period n. The trajectories of the critical points converge to some $p \in P$. Recall from Lemma 2.2.3 that $\varphi_{a,b}(p)$ is a periodic point of period n of the α-multiplication map D_α. We will denote this point by $T(P)$ and we call it the *type of the orbit* P. For a periodic point T of D_α we define the *tongue of type* T as the set of parameters (a,b) for which there exists an attracting periodic orbit of type T. Here we will take a modulo 1 and b in the normal range of $[0, 1]$. If the period of T is n, we will say that the tongue of type T has period n. We want to investigate the order of the tongues as we vary a.

To simplify notation in the following sections we are interested in the α-standard maps for $b = 1$ so we will write $f_{a,1} = f_a$ and $\varphi_{a,1} = \varphi_a$.

Lemma 7.5.1. *The interval joining p with $1/2$ is attracted to p under the iterates of $f_{a,1}^n$, where n is the period of P. Thus, if f_a has an attracting periodic orbit P then $T(P) = \varphi_a(1/2)$.*

Proof. We have that $f'_{a,b}(x) = \alpha + \alpha b \cos(2\pi x)$ and so $f'_{a,b}(1/2) = \alpha - \alpha = 0$ and hence $1/2$ is a critical point of $f_{a,b}$. In particular it is a critical point of f_a.

We now consider the Schwarzian derivative of f_a which we denote by $(\mathcal{S}f_a)(x)$. We have

$$(\mathcal{S}f_a)(x) = \frac{-4\pi^2\alpha\cos(2\pi x)}{\alpha + \alpha\cos(2\pi x)} - \frac{3}{2}\left(\frac{-2\pi\alpha\sin(2\pi x)}{\alpha + \alpha\cos(2\pi x)}\right)^2,$$

which we can simplify to see that

$$(\mathcal{S}f_a)(x) = \frac{2\pi^2(\cos(2\pi x) - 3)}{1 + \cos(2\pi x)}.$$

We note that the numerator is negative for all x and the denominator is positive for all x except when $x = 1/2$. So for all $x \neq 1/2$ the Schwarzian derivative is negative and

$$x \to \frac{1}{2}, \quad (\mathcal{S}f_a)(x) \to -\infty$$

and so the Schwarzian derivative is negative for all x. Hence Singer's theorem applies and as S^1 has no boundary we have that the critical point $1/2$ is in the immediate basin of attraction of p and so we can see the whole interval joining p with $\frac{1}{2}$ is attracted to the periodic orbit P under f_a and so the interval is attracted to p under iterates of f_a^n, where n is the period of P.

To prove the last statement of the lemma we recall the definition of $\varphi_a(x)$

$\varphi_a(x) = \lim_{m \to \infty} \frac{f_a^m(x)}{\alpha^m}$ (mod 1)

which we get from the definition of $\Phi_a(x)$ and the fact that $\Phi_a(x)$ is the lift of $\varphi_a(x)$ to the real line. We also know that $\varphi_a(f_a^n(x)) = \alpha^n \varphi_a(x)$ as φ_a semiconjugates f_a with multiplication by α. So we can see that

$$\alpha^n \varphi_a\left(\frac{1}{2}\right) = \varphi_a\left(f_a^n\left(\frac{1}{2}\right)\right) = \lim_{m \to \infty} \frac{f^{nm}(\frac{1}{2})}{\alpha^m}.$$

We know that $f_a^n(\frac{1}{2})$ is attracted to p and that p is periodic of period n thus we have

$$\lim_{m \to \infty} \frac{f_a^{nm}\left(\frac{1}{2}\right)}{\alpha^m} = \lim_{m \to \infty} \frac{p}{\alpha^m} = \lim_{m \to \infty} \frac{f_a^{nm}(p)}{\alpha^m} = \varphi_a\left(f_a^n(p)\right).$$

Hence using again the fact that φ_a semiconjugates f_a we have that $\varphi_a(f_a^n(p)) = \alpha^n \varphi_a(p)$. Hence $\alpha^n \varphi_a(\frac{1}{2}) = \alpha^n \varphi_a(p)$, which gives that $\varphi_a(\frac{1}{2}) = \varphi_a(p)$. We defined $T(P) = \varphi_a(p)$, which gives the final statement of the lemma; $T(P) = \varphi_a(\frac{1}{2})$. □

Bibliography

[1] R.L. Adler, A.G. Konheim, and M.H. McAndrew. Topological Entropy. *Transactions of the American Mathematical Society*, 114:309–319, 1965.

[2] R.L Adler and M.H. McAndrew. The Entropy of Chebyshev Polynomials. *Transactions of the American Mathematical Society*, 121:236–241, 1966.

[3] V.I. Arnold. Small denominators, I: Mappings of the Circumference onto Itself. *American Mathematical Society Translations*, 46:213–284, 1965.

[4] J. Belair and L. Glass. Universality and self-similarity in the bifurcations of circle maps. *Physica D: Nonlinear Phenomena*, 16:143–154, 1985.

[5] M. Benedicks and A. Rodrigues. Kneading sequences for double standard maps. *Fundamenta Mathematicae*, 206:61–75, 2009.

[6] Keesling J.-Li S. Block, L. and Peterson K. An improved algorithm for computing topological entropy. *Journal of Statistical Physics*, 206:929–939, 1989.

[7] R. Bowen. Entropy for group endomorphisms and homogeneous spaces. *Transactions of the American Mathematical Society*, 153:401–414, 1971.

[8] R. Bowen. Erratum to 'Entropy for group endomorphisms and homogeneous spaces'. *Transactions of the American Mathematical Society*, 181:509–510, 1973.

[9] H. Bruin and S. van Strien. On the structure of isentropes of polynomial maps. *Dynamical Systems*, 28:381–392, 2013.

[10] H. Bruin and S. van Strien. Monotonicity of entropy for real multimodal maps. *Journal of the AMS*, 28:1–61, 2015.

[11] J. Buzzi. Maximal entropy measures for piecewise affine surface homeomorphisms. *Ergodic Theory and Dynamical Systems*, 29:1723–1763, 2009.

[12] Eckmann J.-P. Collet, P. Iterated maps on the interval as dynamical systems. *In Progress in Physics*. Birkhäuser, Boston, MA, 1980.

[13] Y. Young D. Galletly, M. McGuinness and P. Larsen. Arnold tongues in human cardiorespiratory systems. *Chaos*, 14:1–6, 2003.

[14] W. de Melo and S. van Strien. One-dimensional dynamics. *Ergebnisse der Mathematik und ihrer Grenzgebiete(3)*, Springer-Verlag, Berlin, 25.

[15] R.L. Devaney. An Introduction to Chaotic Dynamical Systems. Westview Press, Boulder, Colorado, Second edition, 2003.

[16] A. Dezotti. Connectedness of Arnold tongues for double standard maps. *Proceedings of the American Mathematical Society*, 138:3569–3583, 2010.

[17] E.I. Dinaburg. A correlation between topological entropy and metric entropy. *Doklady Akademii Nauk SSS*, 190:19–22, 1970.

[18] N. Fagella and A. Garijo. The Parameter Planes of $\lambda z^m \exp(z)$ for $m \geq 2$. *Communications in Mathematical Physics*, 273:755–783, 2007.

[19] B. Friedman and C. Tresser. Comb structure in hairy boundaries: some transition problems for circle maps. *Physics Letters A*, 117:15–22, 1986.

[20] G. Levin and G. Świątek. Universality of critical circle covers. *Communications in Mathematical Physics*, 228:371–399, 2002.

[21] G. Levin and S. van Strien. Bounds for maps of an interval with one critical point of inflection type. II. *Inventiones mathematicae*, 2:399–465, 2000.

[22] T. Li and J. Yorke. Period three implies chaosy. *The American Mathematical Monthly*, 82:985–992, 1975.

[23] R. Mañé. Sinks and measure in one-dimensional dynamics. *Communications in Mathematical Physics*, 100:495–524, 1985.

[24] M. Misiurewicz. Horseshoes for mappings of the interval. *The Bulletin of the Polish Academy of Sciences*, 2:167–169, 1979.

[25] M. Misiurewicz. Dynamical Systems. Liguori, Napoli, 1980.

[26] M. Misiurewicz. Chaotic Behaviour of Deterministic Systems. North-Holland, 1983.

[27] M. Misiurewicz and A. Rodrigues. Double Standard Maps. *Communications in Mathematical Physics*, 273:37–65, 2007.

[28] M. Misiurewicz and A. Rodrigues. On the tip of the tongue. *Journal of Fixed Point Theory Applications*, 3:131–141, 2008.

[29] M. Misiurewicz and A. Rodrigues. Non-generic cusps. *Transactions of the American Mathematical Society*, 363:3553–3572, 2011.

[30] M. Misiurewicz and E. Visinescu. Kneading sequences of skew tent maps. *Annales de L' I.H.P.*, 27:125–140, 1991.

[31] C. Robinson. Dynamical Systems: Stability, Symbolic Dynamics and Chaos. CRC Press, 1999.

[32] I.B. Yildiz. Discontinuity of topological entropy for Lozi maps. *Ergodic Theory and Dynamical Systems*, 32:1783–1800, 2012.

[33] J.C. Yoccoz. *Manuscript.*

[34] L.-S. Young. Entropy in Dynamical Systems. eds. Keller and Warnecke, Princeton University Press, pages 313–328, 2003.

Index

Printed in the United States
by Baker & Taylor Publisher Services